TRUST ME:
ISO 42001
AI MANAGEMENT SYSTEM

Gregory Hutchins PE CERM
Margaux Hutchins

The real issue is that when you start to manipulate the information space, you manipulate human behavior.[1]

Eric Schmidt (former CEO Google)

Trust Me: ISO 42001 AI Management System

© 2024 by Gregory Hutchins PE CERM & Margaux Hutchins - VUCAN®, Proactive, Preventive, Predictive, and Preemptive®; CERM: AI risk based, Problem solving | AI risk based, Decision making®, VUCAN®, Architect, Design, Deploy, and Assure® are trademarks of Quality Plus Engineering.

All rights reserved. No part of this book can be reproduced or transmitted in any form or by any means, electronic or mechanical, including photocopying, recording, or by any information storage or retrieval system without written permission from Quality Plus Engineering, except for the inclusion of quotations in a review. No patent or trademark liability is assumed with respect to the use of information contained in this book. Every precaution has been taken in the preparation of **Trust Me: ISO 42001 AI Management System.** The publisher and author assume no responsibility for damages resulting from the use of information herein.

All brand names, trademarks and intellectual property belong to their respective holders. This publication contains the opinions and ideas of its author. It is intended to give helpful and informative material on the subject matter covered. It is sold with the understanding the author and publisher are not engaged in rendering professional services in this book. If the reader requires personal assistance or advice, a competent professional be consulted.

Note: Trust Me: ISO 42001 AI Management System is based on original content. Sections of the book were edited using Chat GPT and Bard.

Disclaimer: The author and publisher specifically refuse any responsibility for any liability, loss, and AI risk, personal or otherwise, which is incurred consequently, directly, or indirectly, of the use and application of any of the contents of this book. If further assistance is required, please seek the assistance of a career, physician, or other healthcare professional.

TABLE OF CONTENTS

Topic	Page
Introduction	7
AI Risk	15
Three Key AI Questions	31
EU AI Act	47
ISO 42001: AI Management System	63
ISO 42001: AI Context	83
ISO 42001: AI Leadership	91
ISO 42001: AI Planning	97
ISO 42001: AI Support	109
ISO 42001: AI Operations	121
ISO 42001: AI Performance Evaluation	129
ISO 42001: AI Systems Improvement	139
Annex Risk - Controls	147
Glossary	207

6   Trust Me: ISO 42001 AI Management System

INTRODUCTION

Trust Me: ISO 42001 AI Management System is a book about the most important AI management system standard: ISO 42001. The ISO 42001 standard is groundbreaking. It is among the first AI standards worldwide.

WHY WRITE THIS BOOK?

AI is transforming how we live and work. AI autonomous decision making is all around us. It is in places we take for granted such as Siri or Alexa.

Understanding AI

It becomes critical we understand and trust this prevalent technology:

> "Artificial intelligence systems have become increasingly prevalent in everyday life and enterprise settings, and they're now often being used to support human decision making. These systems have grown increasingly complex and efficient, and AI holds the promise of uncovering valuable insights across a wide range of applications. But broad adoption of AI systems will require humans to trust their output."[2]

AI systems are inherently risky. Managing and controlling AI risks is central to ISO management systems such as ISO 42001. ISO 42001 provides a framework or lens for identifying, assessing, and managing AI risks.

Global Importance
ISO 42001 is the first AI management system standard. It will have more significance than ISO 9001 as autonomous AI decision making becomes more prevalent.

The authors of the book strive to explain the standard beyond summarizing and paraphrasing the standard. The book breaks down the standard into simple concepts.

WHY READ THIS BOOK?
ISO 42001 is a 10-page standard. Five pages consist of ISO and AI definitions and 30 or so pages consist of Annexes that define risk – controls for each important section of the standard.

Why Write A 200 Page Book For 10 Page Standard?
So, this begs the question: Why write a 270-page book for a 10-page standard?

Here are some ways in which **Trust Me: ISO 42001 AI Management System** adds value to the reader:

- **In-Depth Coverage:** Book provides a comprehensive understanding of ISO 42001 including the underlying theory, complex concepts, and practical applications.

AI Risk 9

- **First ISO 42001 book:** Book is the first comprehensive title on AI management systems.

- **Explain the 'why':** The context and rationale for the standard is explained. Book explains why this standard is important and why specific clauses are critical.

- **Present EU AI Act:** The European Union (EU) Act is the first global AI regulation. The Act requires a quality management system and risk management system. This book explains both requirements.

- **Mapped 1-1 to ISO 42001:** Chapters in the book are mapped to ISO 42001.

- **Authoritative source:** Greg Hutchins is an expert on ISO 31000, risk based auditing, and AI. Greg is the author of the best-selling **ISO 31000: ERM** (4.9/5.0 Amazon reviews), **Trust Me: AI Risk Management**, **Value Added Auditing**, and other risk management and assurance books.

- **Comprehensive:** Book covers ISO 42001 from social, technical, assurance, and risk management perspectives.

- **Great investment:** ISO 42001 standard in the U.S. and EU costs about $220. This book provides a comprehensive understanding of the standard rather than purchasing the standard and reading its turgid requirements.

- **Improve risk based thinking; risk based, problem solving; and risk based, decision making:** Book introduces critical risk concepts that help readers understand and operationalize risk - controls to make better risk based, problem solving and decision making.

- **Simple and readable:** Reading ISO 42001 requires focus and concentration. The author writes in a simple and straight forward fashion explaining difficult AI and ISO concepts.

- **Reference Guide:** The book's Glossary explains 100's of quality, risk, and AI terms.

TRUST ME SERIES OF BOOKS

This is the second book in the AI trust series. AI trust is an important topic.

Inputs and Outputs of AI Systems

AI is not a new technology. AI decision making systems have been used for more than 50 years starting with Lisp programming and other tools. Neural nets are an early form of generative models and have been used for many years.

The difference today is the computing, storage, and performance of the neural nets are millions of times more efficient, effective, and risky than in previous AI systems.

But What Happens If AI Is Not Trusted?

If the trust factor is not secured, then bad things can happen:

> "But if consumers do not trust AI, adoption may be limited to less-consequential tasks, such as recommendations on streaming services or contacting a call center in the search for a human."[3]

AI Risk

The One Trust Question

Can AI trust be summarized into one question?

> "The highest bar for AI trust can be summed up in the following question: What would it take for you to trust an AI system with your life? ... We sort trust in an AI system into three main categories.
>
> - Trust in the performance of your AI/machine learning model.
>
> - Trust in the operations of your AI system.
>
> - Trust in the ethics of your workflow, both to design the AI system and how it is used to inform your business process."[4]

AI In Healthcare Decision Making

All humans go to a doctor. Healthcare autonomous or AI decision making can be both consequential and personal:

> "Artificial intelligence (AI) can transform health care practices with its increasing ability to translate the uncertainty and complexity in data into actionable—though imperfect—clinical decisions or suggestions. In the evolving relationship between humans and AI, trust is the one mechanism that shapes clinicians' use and adoption of AI.
>
> Trust is a psychological mechanism to deal with the uncertainty between what is known and unknown. Several research studies have highlighted the need for improving AI-based systems and enhancing their capabilities to help

clinicians. However, assessing the magnitude and impact of human trust on AI technology demands substantial attention. Will a clinician trust an AI-based system? What are the factors that influence human trust in AI? Can trust in AI be optimized to improve decision making processes?"[5]

AI In Legal Decision Making

Lawyers do not understand AI so the profession is taking a go slow approach. Lawyers are scared because AI is redefining their profession:

"In a recent survey of lawyers conducted by the Thomson Reuters Institute, 82% of legal professionals said they believe that generative AI such as ChatGPT *can* be applied to legal work, with 59% of partners and managing partners sharing generative AI *should* be applied to legal work."[6]

Trust <=> AI Risk

Public policy has the ability to govern AI. Public policy in AI technology is running ahead of the ability of humans and policy makers to develop laws, regulations, and specifications. Humans need to define the 'rules of engagement' between humans and machines.

FINAL THOUGHTS

- AI is a disruptive technology.

- ISO 42001 is only 10 pages long but is a highly influential global AI standard.

AI Risk

- ISO 42001 is critical for establishing trust in the design, development, and deployment of AI systems.

- **Important distinction:** An important clarification needs to be made. ISO 42001 is an AI Management System that is abbreviated to AIMS. AI is also a social technical system. So, many call it an 'AI system.' In this book, AIMS refers to the ISO 42001 standard. 'AI system' refers to AI context, technology factors, social factors, apps, and products.

14   Trust Me: ISO 42001 AI Management System

AI RISK

AI is around us making important decisions that impact lives and work. For example:

> "Every time you speak to a virtual assistant on your smartphone, you are talking to an artificial intelligence — an AI that can, for example, learn your taste in music and make song recommendations that improve based on your interactions.
>
> However, AI also assists us with more risk-fraught activities, such as helping doctors diagnose cancer. These are two very different scenarios, but the same issue permeates both: How do we humans decide whether or not to trust a machine's recommendations?"[7]

WHAT IS AI?

There are many definitions of AI. For much of this book, we focus on AI decision making specifically risk based, problem solving and risk based, decision making.

What Is AI?

In terms of these two elements, we define AI as the ability of a machine to resemble human intelligence to do tasks, solve problems, make decisions, and repeatedly improve and learn.

16   Trust Me: ISO 42001 AI Management System

This definition has many human qualities to it. And this is what makes AI fearful. What decisions will AI make that impacts humans? How much will AI decisions impact humans and in what ways? And often, these public facing decisions are not explainable with no human in the middle to explain the decision.

AI Is Big - Important As Steam Engine

AI is probably the most consequential technology since the invention of electricity and steam engine. Why?

> "Artificial intelligence (AI) technologies and frameworks could radically boost efficiency and productivity in nearly every field. It can enable better, faster analysis of imagery in fields ranging from medicine to national security. [8]

What Is Explainable AI?

Explainable AI has become a necessary and sufficient condition for trustworthy AI. It is well known generative AI is prone to hallucinations. The challenge is that developers and users sometimes cannot explain how the system reached its decisions and conclusions based on its data sets and processes.

The following quotation addresses the importance of explainable AI:

> "Explainable artificial intelligence refers to methods and techniques that produce accurate, explainable models of why and how an AI algorithm arrives at a specific decision so that AI solution results can be understood by humans. Without explanations behind an AI model's internal

functionalities and the decisions it makes, there is a risk that the model would not be considered trustworthy or legitimate."[9]

What is AI Risk?

Let us first define risk. Risk is the "effect of uncertainty on objectives." This is the ISO 31000 definition of risk. This is a little difficult to operationalize because it is difficult or even impossible for uncertainty to impact an objective. An objective is immutable.

So, we say risk is the "effect of uncertainty on *the ability to achieve* an objective." As you can see, we added the 'ability to achieve' to the definition. Uncertainty impacts the achievement to reach the objective, not the objective.

Another way to think about how uncertainty impacts the achievement of the objective is to say AI risks arise from the possibility of future events that could have a negative impact on a company's ability to achieve its AI objectives.

AI RISKS

AI systems and apps are used more often as automated or autonomous decision making.

Risky AI Decision Making

In the early days of AI, the three main elements of a decision model were known. The inputs of the system were known. The process (engine) of the system was logical. And the outputs were understood. In other words, the inputs and outputs were causal. As the

18   Trust Me: ISO 42001 AI Management System

decision making systems became sophisticated, the inputs and outputs of the process were still correlated.

Generative AI systems now hallucinate. Inputs are often not known. The decision making process is a black box. The outputs are not causal, not correlative, and often unpredictable, unexplainable, and hallucinatory. This is a problem.

ChatGPT which is often unexplainable and unpredictable exemplifies the problem:

> "Trust is grounded in predictability. It depends on your ability to anticipate the behavior of others. If you trust someone and they do not do what you expect, then your perception of their trustworthiness diminishes."[10]

AI Impacts People

Fear of losing jobs due to AI is a major personal fear of many. For example, CEO Suumit Shah in India recently decided to cut staff:

> "The worst nightmare of employees - being eventually replaced by AI, may well have arrived, a bit sooner than we expected. ... We had to lay off 90% of our support team because of this AI chatbot. Tough? Yes. Necessary? Absolutely."[11]

AI Risk Impacts

Risks arise if there is no human in the middle of the decision making or problem solving process. These social and technical risks can arise:

AI Risk

- **Workers:** AI is already impacting employment, housing, and other decisions. What will people do in a world without work.

- **Professions:** AI is impacting the Future of Work from accounting, app development, auditing, and legal.

- **Companies:** AI is integrated into many products and apps. Development requires internal risk protocols.

- **Healthcare access:** Could or would an AI system deny a person healthcare if a human had a pre-existing condition?

- **Tenant screening:** Could an AI system refuse a human access to an apartment due to a low credit card score, ethnicity, or critical record?

- **Criminal justice:** Currently in the U.S., AI systems recommend to judges criminal penalties and incarceration based on the history of the perpetrator. Could the AI be wrong or biased?

- **Facial recognition:** Facial recognition can be used to identify persons of interest and others for special or adverse treatment. Adverse treatment would deny personal access to a hospital or an airplane.

- **Employment screening:** Currently in the U.S., AI does initial screening of job applicants. It is essential this process is fair and transparent. New York Law 144 requires risk based audits of autonomous hiring systems.

- **Business and work automation:** AI is used in many business functions such as automating work. What do people then do with their dreams, time, and make a living. Tough questions that have not been addressed.

- **Homelessness:** Homelessness and mental challenges are rampant in many parts of the world. What happens to these humans when there are few or no resources due to the AI decision making.

- **Future issues:** Risk statutes, regulations, definitions, assurance, accountability, and implementation are evolving and multiplying quickly. Each brings its own uncertainty.

AI Existential Risk

AI is often depicted as a Terminator like threat. Sometimes, this doomsday scenario is called an existential or singularity threat to humanity. The threats could be an AI superintelligence that becomes uncontrollable in its decision making and innovation. Another scenario is the creeping technology fear where AI gradually worms into critical systems such as nuclear power plants and gradually disrupts and corrupts them.

AI Fear In Numbers

As can be seen by the above threats, many fear and do not trust AI. The numbers are stunning:

> "Most people (85 percent) believe AI will deliver a range of benefits, including efficiency, effectiveness, innovation and resources."[12] "Alongside the benefits, most people (73 percent) perceive significant risks in AI."[13]

Paper Clip Analogy Of Human Destruction

The paper clip analogy is a wry example of the existential threat:

AI Risk

"A lot has been written about AI's as existential risk. The worry is that they (AI) will have a goal, and they will work to achieve it even if it harms humans in the process. You may have read about the 'paperclip maximizer': an AI that has been programmed to make as many paper clips as possible, and ends up destroying the earth to achieve those ends. It is a weird fear."[14]

ISO 31000 RISK MANAGEMENT

ISO 31000 is another document that impacts the AIMS. ISO 31000 is the general risk management standard that can be applied with and integrated into ISO 42001.

ISO 31000

By following the ISO AI risk management process, companies identify, assess, and manage AI risks, thereby increasing the likelihood of achieving their objectives and improving their AI performance.

ISO 31000 provides guidelines for managing risks in companies, regardless of their size, sector, or location. ISO 31000 is not an AI standard or guideline. ISO 31000 describes a process that can be used for managing AI risks.

ISO 31000 Elements

ISO 31000 can be tailored to analyze AI risks, specifically to:

22 Trust Me: ISO 42001 AI Management System

- **Identify AI risks:** This involves recognizing and documenting potential risks that could affect achieving the company's AI objectives.

- **Analyze AI risks:** This involves assessing the likelihood and potential consequence of each identified AI risk.

- **Evaluate AI risks:** This involves comparing the analyzed AI risks against the company's criteria to determine which risks need to be addressed.

- **Treat AI risks:** This involves developing and applying strategies to address the identified AI risks. This is like risk management.

- **Monitor and review AI risks:** This involves monitoring AI risks and implementing risk management strategies.

ISO 31000 is beyond the scope of this book. However, purchase **ISO 31000: Enterprise Risk Management** book on Amazon. It has 4.9 stars out of 5 and is the best-selling ISO risk book.

RULES OF ENGAGEMENT

As we discussed, AI is a social and technical system. It becomes important to develop rules of engagement between humans and machines.

Military Rules of Engagement

The expression 'rules of engagement' usually has a military and business meaning depending on the expressions use and context.

AI Risk

In the military context, 'rules of engagement' are the boundary conditions, taxonomy, governance, principles, and guidelines that dictate how opposing adversaries conduct war.

In acts of war, the rules of engagement also dictate how to comply with international law, align with strategic objectives, and assure military actions are ethical, controlled, and defendable.

Business Rules of Engagement

In the business context, 'rules of engagement' are the guidelines how business is conducted between partners, stakeholders, and teams. Often, lawyers define these business rules of engagement. The rules are intended to prevent conflicts, comply with regulations, ensure understanding on common goals, ensure communications, and even define authorities and responsibilities.

Technical Rules Of Engagement

Many are worried AI systems will result in existential and catastrophic risks for humans. Just think if a military AI system has access to nuclear codes and could make an independent decision to launch nuclear missiles if the system perceived a military threat, an adversary preemptive launch, or other threat signatures.

Social Rules Of Engagement

AI also has social rules and impacts. So, the following are considered in developing social engagement rules:

- **Ethical guidelines:** These include minimizing bias, ensuring transparency, and avoiding harm.

- **Human oversight:** Humans need to be in the middle and partner in autonomous decision making involving housing, healthcare, and jobs.

- **AI explainability:** As AI becomes more powerful, AI generative systems can hallucinate. This means they are not explainable. Human intervention, explanation, and mediation will become the norm between public facing AI systems and those impacted.

AI Rules of Engagement

So, what are the rules of engagement between humans and machine? And, who defines them? The machine is robotics and AI. You can think of robotics as the body in terms of motion, articulation, and manipulation that mimics humans. You can think of AI as the head in terms of autonomous decision making and problem solving.

Here is a hard reality. The world does not have 'rules of engagement' between AI and humans. The challenge is how to work with and interact with robots and AI. If humans do not get this right, this is going to be very risky to humans. Why? AI and robotics are growing exponentially in importance, prevalence, and human risk.

Defining AI Rules Of Engagement

Autonomous decision making is the shift from human to AI decision making. The risks of using and developing AI will require more caution. New rules for autonomous problem solving and decision making will have to be developed.

AI Risk 25

There are lots of implications with this. What will be the rules of AI engagement look like? What are they? How will they be implemented? Will AI replace work, careers, and jobs? Will AI augment the work of human? What will the augmentation and collaboration look like, in what areas, and how will joint decisions be made? Lots of questions. Lots of risks. Few answers.

Real World Scenarios

Just consider these real-world scenarios:

- "How does a software engineer gauge the trustworthiness of automated code generation tools to co-write functional, quality code?

- How does a doctor gauge the trustworthiness of predictive healthcare applications to co-diagnose patient conditions?

- How does a warfighter gauge the trustworthiness of computer-vision enabled threat intelligence to co-detect adversaries?"[15]

Human In The Loop

Many approaches are being developed to define AI trust. One is called human in the loop. In this decision making, AI does the analysis to crunch data, solve problems, and propose solutions. Humans provide the governance, assurance, and make the final decision.

The U.S. military follows this method:

"One way to reduce uncertainty and boost trust is to ensure people are in on the decisions AI systems make. This is the approach taken by the U.S. Department of Defense, which requires that for all AI decision making, a human must be either in the loop or on the loop. In the loop means the AI system makes a recommendation but a human is required to initiate an action. On the loop means that while an AI system can initiate an action on its own, a human monitor can interrupt or alter it."[16]

ISO 42001 Defines AI Rules of Engagement

The European Union (EU) AI Act has set up legal 'rules of engagement' with AI. The challenge is how these rules will be defined and operationalized. In other words, how will they be implemented?

ISO 42001 is a starting point for companies to start defining and implementing the management system rules of AI engagement.

AI RISK BASED, DECISION MAKING

Risk is the lens for looking at and understanding AI. Just consider the following:

> "In contrast, an AI can't rationalize its decision making. You can't look under the hood of the self-driving vehicle at its trillions of parameters to explain why it made the decision that it did. AI fails the predictive requirement for trust."[17]

AI Is Not Going Away!

So, trust and rules are very important in today's world of uncertainty. Why? AI is not going away.

AI Risk

"Today, the amount of data that is generated, by both humans and machines, far outpaces humans ability to absorb, interpret, and make complex decisions based on the data. Artificial intelligence forms a basis for all computer learning and is the future of all complex decision making."[18]

Risk based, problem solving and Decision Making

The concept of managing and controlling AI risks is central to ISO AI management systems, since it provides a framework for identifying, assessing, and managing AI risks. Companies can then make informed risk decisions, prevent problems, and seize opportunities.

We call this 'risk based, problem solving and risk based, decision making.' These concepts are not part of ISO 42001, but are critical elements of our approach to enterprise risk management (ERM) of AI systems.

In Chat GPT, the inputs into the process are often based on very large language models. The input data can be biased or simply corrupt. In other words, the inputs, process and outputs of AI problem solving, and decision making are unknown and risky. Hence, AI decision making must evolve to be risk based, problem solving and risk based, decision making.

Risk Based, Decision Making Challenges

Risk is the lens for using and developing AI systems. However, many elements of generative AI systems are unknown and unexplainable.

The following are a few conclusions experts have reached regarding developing AI systems:

- Risk based, problem solving and decision making will dominate the new world of AI. There will be more questions than answers. Some key risk based, decision making questions will arise:

    o How is the AI system going to decide?

    o What is to be decided by the AI system?

    o What are the social and technical factors of the AI systems?

    o Who is the human in the middle of the AI decision making?

    o What are the qualifications of the human in the middle?

    o What are the rules of human and machine engagement for risk based, problem solving and decision making?

    o What data was used to train the model?

    o Is any of the data used in the model protected?

    o What are the risks in the AI system?

    o How are the risks mitigated?

FINAL THOUGHTS

- ISO 42001 can be thought as a guide to defining the rules of engagement between humans and machine. Hence,

AI Risk 29

Trust Me: ISO 42001 AI Management System was developed.

- AI is a social and technical system.

- Social systems address ethics, trust, and explainability.

- Technical systems address infrastructure, data sets, use cases, and algorithms.

- Autonomous AI decision making requires trust.

- Importance of trust is captured in below quote:
  - "Trust is the currency of the AI era, yet, as it stands, our innovation account is dangerously overdrawn. Companies must move beyond the mere mechanics of AI to address its true cost and value - the 'why' and 'for whom.'[19]

- AI impacts people in often unknown and unexplainable ways.

- AI is the ability of a machine to mimic human reasoning and behaviors.

- AI because of movies such as Terminator results in the fear of existential AI threats.

- AI decision making and problem solving must be reasonable and explainable since they impact people in unknown ways.

- AI rules of engagement is how humans and machine collaborate.
- 'Human in the loop' is one method or rule for human and machine collaboration.

THREE KEY AI QUESTIONS

Regulators around the world want safe, trustworthy, and explainable AI systems.

THREE KEY QUESTIONS

There is a fine balance that global regulators must have between encouraging innovation, while providing the requisite guard rails to this new technology:

> "Regulators are also looking for AI to have a net positive impact on society, and they have begun to develop enforcement mechanisms for human protections, freedoms and overall well-being. Ultimately, to be accepted by users — both internally and externally — AI systems must be understandable, meaning their decision framework can be explained and validated. They must also be resolutely secure, even in the face of ever-evolving threats."[20]

Three Key Questions

Three key questions for AI trust and assurance arise that are discussed in this chapter:

1. What standards will AI comply with?
2. How will AI be audited and assured?
3. Who conducts AI audits?

WHAT STANDARDS WILL AI COMPLY WITH?

The number of AI laws and regulation are expanding quickly. The number of regulations at the state level impacting AI is about 50 according to recent news reports. This will increase in the future as more AI decision making impacts humans.

AI Regulations

AI regulations and standards are coming. They may be global standards and guidelines such as ISO 42001 or they may be sector specific. The challenge is there is more awareness of the dangers of neglect or inaction to develop AI standards:

> "Unfortunately, not all AI is built to the standards of driverless cars. Because cars are especially destructive and dangerous, the automotive industry is well regulated. Now we are coming to realize that software systems, with or without AI, should also be well regulated. We now know that extraordinary damage can be done when, for example, foreign governments interfere in close elections through social media."[21]

U.S. AI BILL OF RIGHTS

The United States is behind the European Union in developing statutes, rules, and regulations for high risk AI systems.

Bill Of Rights Principles

U.S. AI Bill of Rights was developed in 2022 out of the White House. It is not a law. It is not a regulation. It is a set of guiding principles.

It is a set of social and technical principles, specifically:

- **Safe and effective AI systems:** AI systems should be architected, designed, and deployed so they are reliable (technical elements) and minimize harms (social elements).

- **Algorithmic discrimination protections:** AI systems should be deployed so they do not discriminate or make biased decisions (technical elements).

- **Data privacy:** Personal identity information and privacy are protected in deploying the AI systems (social elements).

- **Notice and explanation:** AI systems should be transparent and explainable in their decision making (technical and social elements).

- **Human alternatives, consideration, and fall back:** Companies and developers of the AI systems should be identifiable and be accountable for the systems (social elements).

ARTIFICIAL INTELLIGENCE RISK MANAGEMENT FRAMEWORK (AI RMF)

Early in 2023, NIST developed the Artificial Intelligence Risk Management Framework commonly abbreviated AI RMF. U.S. National Institute of Standards and Technology (NIST) is part of the Department of Commerce. This framework is a guideline and is evolving into a hub of AI standards development in the United States.

AI RMF Structure

The NIST RMF follows the same structure as the cyber security framework commonly called the CSF. The CSF was released in 2014. The CSF is also linked to the privacy framework which was released in 2020. This also follows the same model.

These 'living documents' are voluntary, rights preserving, risk based, and use agnostic. The AI RMF is organized into core functions, subcategories, and implementation profiles. Another common feature of most AI frameworks is they are risk based and can be used in different decision making contexts.

Trust Manifesto

ARM AI, semiconductor company, has a Trust Manifesto involving:

1. "We believe all AI systems should employ state-of-the-art security.

2. Every effort should be made to eliminate discriminatory bias in designing and developing AI decision systems.

3. We believe AI should be capable of explaining as much as possible.

4. We urge further effort to develop technological approaches to help AI systems record and explain their results.

5. Users of AI systems have a right to know who is responsible for the consequences of AI decision making.

6. Human safety must be the primary consideration in the design of any AI system

7. We will support efforts to retrain people from all backgrounds to develop the skills needed for an AI world."[22]

EU AI ACT

EU AI Act is a risk-based system for AI assurance and accountability. The EU AI Act was approved in 2024. The Act is important for many reasons. It shapes Europe's AI, digital, and privacy futures. It is the first law regulating the design and deployment of AI from a social and technical risk perspective.

What is in the EU AI Act. The Act looks at AI or autonomous decision making through a risk lens. This means the Act requires system developers to assess the risks of their AI systems. ISO 42001 compliance is critical since it offers a management systems approach to assess AI risks.

AI Act Levels of Risk

The EU AI Act identifies 4 levels of risk:

1. **Unacceptable risk:** These systems are direct and substantial threats to public safety and people. The Act prohibits AI that exploits people vulnerabilities such as ethnicity, age, or disabilities.

2. **High risk:** These systems include public infrastructure, enforcement, and employment. The EU AI Act identifies 8 areas as high AI risk:

    a. Biometric identification and categorization of natural persons.

b. Management and operation of critical infrastructure.

c. Education and vocational training.

d. Employment, worker management and access to self-employment.

e. Access to and enjoyment of essential private services and public services and benefits.

f. Law enforcement.

g. Migration, asylum and border control management.

h. Assistance in legal interpretation and application of the law.

3. **Limited risk:** These systems include generative AI such as LLM's and chatbots.

4. **Limited risk:** These systems include video games and spam filters.

HOW WILL AI BE AUDITED AND ASSURED?

The second key question about AI safety is: How will AI be audited and assured? Or in other works, how do humans know AI will be rights preserving and safe? Great question. But, difficult to answer.

The EU, Canada, Japan and many countries intend to use ISO 42001 and conformity assessment to audit, assure, and make AI accountable.

Conformity Assessment

Conformity assessment is a process to make sure an AI product, app, or service can meet or satisfy standards, regulations, or specifications. Think of conformity as the technical methods or means to establish trust in an app, service, or software.

The goal of conformity assessment is to establish trust and confidence. Conformity assessment is used to evaluate products and services. So, what are the steps the EU is considering for AI conformity assessment?

EU AIA AI Assurance

At a high level, the EU AI Act proposes the following:

- **Step 1:** High risk AI system or app is developed.

- **Step 2:** High risk system undergoes conformity assessment and complies with AI requirements. For some high risk systems, a notified body in involved in the conformity assessment.

- **Step 3:** Registration of the standalone ISO 42001 system is placed in an EU database.

- **Step 4:** Declaration of conformity needs to be signed and the AI systems has the CE mark before it can be placed on the market.[23]

Conformity Assessment Examples

AI conformity assessment is often achieved in different industries:

- **Manufacturing:** Company achieves conformity with ISO 9001:2015 in manufacturing by applying a quality management system that includes procedures for product design, and production.

- **Construction:** Company in the construction industry a company achieves conformity with ISO 14001:2015 by applying an environmental management system that includes procedures for environmental impact assessment, waste management, and pollution prevention.

- **Healthcare:** Company in the healthcare industry achieves conformity with ISO 45001:2018 by applying an occupational health and safety management system that includes procedures for hazard identification and risk assessment, employee training, and accident investigation.

- **Financial services:** Company in the financial services industry achieves conformity with ISO 27001:2013 by applying an information security management system that includes procedures for access control, data encryption, and incident response.

In the same way by complying with ISO 42001, companies achieve conformity with applicable requirements and improve their AI performance.

How Does Conformity Assessment Provide Trust?

Conformity assessment indicates trust in AI by:

- **Providing confidence:** Once a product or service has been evaluated, app users, regulators, and stakeholders

Three Key AI Questions

should be confident the app is fit for service and fit for purpose based on complying with a standard, guideline, or regulation such as ISO 42001.

- **Assuring consistency:** A hallmark of quality, safety, and trust is consistency through the application of an international standard such as ISO 42001, specifically consistent architecture, design, deployment assurance, and use of the AI system.

- **Proving safe:** Once an app is conformity assessed, then it is assumed it can be used safely for the context and purpose for which it was designed and deployed.

Certification Body Approach to Auditing

Global certification bodies are already auditing and certifying companies and government agencies to ISO 42001 because AI has become a major public safety issue.

For example, SGS is a global certification body and promotes third party auditing (conformity assessment):

"ISO/IEC 42001 certification follows successful completion of an audit and enables you to:

- Implement AI safely, with evidence of responsibility and accountability.

- Consider security, safety, fairness, transparency and data and AI system quality throughout the lifecycle.

- Show that introducing AI is a strategic decision with clear objectives.

- Indicate strong governance concerning AI.

- Strike a balance between governance and innovation.

- Ensure AI is used responsibly, especially concerning its continuous learning.

- Ensure safeguards are in place.

- Combine key frameworks with experience to implement crucial processes like risk, lifecycle and data quality management."[24]

Conformity Assessment Methods

Conformity assessment involves different methods:

- **Testing:** App testing is a common method to determine if an app meets a technical specification.

- **Product mark:** The CE mark is often found in electronic consumer products. The EU CE mark on a product indicates it meets the EU's health, safety, and environmental requirements.

- **Inspection:** Visual inspection, test, or measurements are often used to certify and ensure compliance.

- **Certification:** Certification auditing is done by a third party, independent auditor. Certification auditing is a formal declaration or certificate that a product, system, or service

Three Key AI Questions   41

complies with a regulation or standard. ISO 9001 certification is probably the most recognized form of certification.

- **Risk-based auditing:** Most conformity assessment audits are risk based. Auditing is a systematic examination and review of policies, procedures, work instructions related to a specification or standard. The purpose of the audit is to assure applicable standards are being met and organizational objectives are being met.

- **Applying the ISO 42001 AIMS as best practice:** A company does this to improve its AI development operations and processes.

- **Self – certification:** Self-declaring ISO 42001 conformity means an organization attests it complies with the requirements of a management system standard such as ISO 42001, conducts a compliance audit, and self attests all conditions are met.

OECD Product Trust Label

Companies and NGOs are developing AI product labels such as OECD's Trust Label:

> "The AI Trust Standard & Label describes the characteristics of an AI product with regards to: Transparency, Accountability, Privacy, Fairness and Reliability. This tool provides a way to describe ethically relevant characteristics of AI systems in a testable and enforceable way, with a focus on transparency, fairness, accountability, privacy and robustness."[25]

Lifecycle AI Audits

The approach a company takes to achieve AI conformity using ISO 42001 depends on the size and complexity of the company, type of apps and services, risk, and purpose of use.

For example, public facing, high risk AI will be assessed at a deeper and broader level than a low risk gaming app. This captures the complexity of assessing high risk apps:

> "One approach is to scrutinize the potential for harm or bias *before* any AI system is deployed. This type of audit could be done by independent entities rather than companies, since companies have a vested interest in expedited review to deploy their technology quickly. ... You basically have to audit algorithms at every step of the way to make sure that they don't have these problems. It starts from data collection and goes all the way to the end, making sure that there are no feedback loops that can emerge out your algorithms. It's really an end to end endeavor."[26]

WHO CONDUCTS THE AI AUDITS?

The third question is: who conduct the AI audits? ISO 42001 audits are conducted by authorized companies and notified bodies such as SGS, TUV, and others. However, the question of what are the knowledge, skills, and abilities of the management system auditors still has not been decided. Below are several options regarding auditing of AI systems.

Internal Auditing

The question of who audits AI systems will come down to risk, context, needs, regulatory compliance, and fitness for purpose. In general, public facing, high risk AI systems will require higher levels of assurance and accountability than low risk AI systems.

Internal audit teams must have audit experts, domain experts, data scientists, and AI engineers. Domain experts bring deep knowledge, skills, and abilities to the audit team. AI engineers have critical information on how the AI system is architected, designed, deployed, and assured. As well, AI engineers test, verify, and validate the training data used by the AI system.

ISO 42001 AIMS Auditing

The AIMS auditor team needs experience in management systems auditing including ISO 19011. ISO 19011 is the ISO standard for auditing management systems.

AIMS auditing would normally be conducted by a lead ISO assessor. The challenge with AIMS auditing is AI is a social technical system that is technically deep and socially broad in its impacts. AI systems may be difficult to audit as well because of their technical complexity and black box characteristics.

Black Box Problem

Auditors require an audit trail of information. What makes a good AI audit? AI inputs are well understood with high quality data. AI process is well defined with recognizable. AI problem solving and decision making process is explainable. The AI outputs are clear and logical with little bias or errors. The reality is that generative AI seldom meets these conditions.

AI Auditing Challenges

Auditing AI systems is risky for the following reasons:

- **Data quality:** Data input data may be biased and low quality.

- **Box box process:** AI process may not be understood or explainable.

- **AI decision making process:** AI process may be opaque, not transparent, and unexplainable.

- **Lack of standardization:** AI is a new technology with few standards. ISO 42001 is relatively new and applicable for low risk AI applications.

- **Cost of risk based Audits:** AI audits of complex systems are time consuming and very expensive.

- **Management system audits:** Management system audits can be checking the box exercise instead of focusing on AI safety and preserving human rights.

- **Control objective:** ISO 42001 is based on auditing control objectives. This provides relatively low-level assurance.

FINAL THOUGHTS

- ISO 42001 is the first AI management system (AIMS). It is like ISO 9001, but for Artificial Intelligence.

- ISO 42001 consists of 5 pages of definitions, 10 pages of the standard, and more than 30 pages of risk – controls.

Three Key AI Questions 45

- ISO 42001 has many 'shall' requirements. In this book, we do not use 'shalls' but use the word 'require' or 'need to' be implemented.

- In the U.S., there is an exponential increase in AI standards and statutes.

- In the EU, the AI act is the principal regulation for AI.

- The EU AI Act may become the global benchmark for AI regulation.

- The EU AI Act does not define how to comply with quality management system or risk management system requirements. However, the EU Act is updated regularly where QMS and RMS may be defined more clearly in the future.

- AI Act specifies higher risk AI systems must have written or otherwise documented risk - controls.

- In the U.S., national AI statutes are not coming out of Congress.

- Most AI statutes are coming from states and local jurisdictions.

- The NIST Artificial Intelligence Risk Management Framework (AI RMF) was developed in 2023. It is becoming a pivot point for AI guidelines and standards in the U.S.

- NIST AI RMF is divided in to 4 elements: 1. Govern; 2. Manage; 3. Map; and 4. Measure.

EU AI ACT

This chapter covers AI regulations in the EU and U.S. Many regulations have not been deployed or enforced. As well, these regulations, standards, and guidelines change frequently based on the maturity and implementation of high risk AI systems.

EU AI ACT QUALITY MANAGEMENT SYSTEM REQUIREMENTS

While the exact language of the EU AI Act is frequently rewritten, the general outline and interpretation are shown below.

Quality Management System Requirement

The EU AI Act requires ISO 9001 Quality Management System (QMS) for High Risk AI systems according to Article 17 of the Act:

> "Providers of high risk AI systems shall have a quality management system that ensures compliance with this Regulation. It shall be documented in a systematic and orderly manner in the form of written policies, procedures or instructions and can be incorporated into an existing quality management system under Union sectoral legislative acts."[27]

Interpretation: High risk AI systems must have a quality management system such as ISO 9001. The systems must be documented consisting of policies, procedures and instructions that impact the systems architecture, design, deployment, and assurance.

The required elements of the quality management system include:

Quality Management System

1. "Providers of high risk AI systems shall put a quality management system in place that ensures compliance with this regulation. That system shall be documented in a systematic and orderly manner in the form of written policies, procedures and instructions, and shall include at least the following aspects:"

Interpretation: ISO 9001 is the quality management system standard. ISO 9001 must be tailored to the AI system's risk, use cases, and context. The following are minimum requirements for an AI system:

a. "A strategy for regulatory compliance, including compliance with conformity assessment procedures and procedures for the management of modifications to the high risk AI system;"

Interpretation: Conformity assessment includes product marks such as an AI watermark and management system assessment. Changes in the system require further risk evaluation.

EU AI Act 49

b. "techniques, procedures and systematic actions to be used for the design, design control and design verification of the high risk AI system;"

Interpretation: Policies, procedures, and instructions are developed and deployed for the design, control, verification, and validation of high risk AI systems.

c. "techniques, procedures and systematic actions to be used for the development, quality control and quality assurance of the high risk AI system;"

Interpretation: Quality management, quality assurance and quality control are required in the development of high risk AI systems. These must be documented with AI policies, procedures, and work instructions.

d. "examination, test and validation procedures to be carried out before, during and after the development of the high risk AI system, and the frequency with which they have to be carried out;"

Interpretation: The quality management system of high risk AI systems requires examination, test, and validation throughout its development lifecycle which include deployment and retirement.

e. "technical specifications, including standards, to be applied and, where the relevant harmonised standards are not applied in full, or do not cover all of the relevant requirements set out in Chapter II of this

Title, the means to be used to ensure that the high risk AI system complies with those requirements;"

Interpretation: AI specifications are technical and social. Technical standards involve development and deployment. Social standards involve trust, equity, and explainability. If standards are not used in full, not developed, or not adhered to, the scope of the compliance must be redefined.

f. "systems and procedures for data management, including data acquisition, data collection, data analysis, data labelling, data storage, data filtration, data mining, data aggregation, data retention and any other operation regarding the data that is performed before and for the purposes of the placing on the market or putting into service of high risk AI systems;"

Interpretation: Data used in the AI system should be identified and managed appropriately throughout its lifecycle.

g. "the risk management system referred to in Article 9;"

Interpretation: Risk management systems should be designed, deployed, and assured according to Article 9 of the EU AI Act.

EU AI Act

h. "the setting-up, implementation and maintenance of a post-market monitoring system, in accordance with Article 61;"

Interpretation: Article 61 specifies the 'Post Market Monitoring by Providers and Post Market Monitoring Plan for High risk AI Systems.' AI systems designers must establish and document a system for checking systems that have been deployed to assure they meet the performance and risk criteria of the AI Act regulation.

i. "procedures related to the reporting of a serious incident in accordance with Article 62;"

Interpretation: Article 62 specifies 'Reporting of Serious Incidents and of Malfunctioning.' Developers of high risk systems must report problems or breaches of the AI systems that have been placed into service.

j. "the handling of communication with national competent authorities, other relevant authorities, including those providing or supporting the access to data, notified bodies, other operators, customers or other interested parties;"

Interpretation: Communications protocols are developed with Certification Bodies, interested parties regarding the safety of high risk AI systems.

k. "systems and procedures for record keeping of documentation and information;

Interpretation: Procedures and instructions are developed for record keeping of high risk systems.

l. "resource management, including security of supply related measures;"

Interpretation: Adequate resources are available and deployed for in house and supplied materials and services used in the systems.

m. "an accountability framework setting out the responsibilities of the management and other staff with regard to all aspects listed in this paragraph."

Interpretation: Responsibility and authority actions and protocols of the system are defined and deployed throughout the AI app's lifecycle.

Implementation of QMS

2. "The implementation of aspects referred to in paragraph 1 shall be proportionate to the size of the provider's organisation. Providers shall in any event respect the degree of rigour and the level of protection required to ensure compliance of their AI systems with this Regulation."

Interpretation: The implementation of the requirements in the previous paragraph are proportionate to the size, use and context of the systems developer and the risks of the system. High risk AI systems require more assurance and risk - controls.

EU AI Act

a. "For providers of high risk AI systems that are subject to obligations regarding quality management systems or their equivalent function under relevant sectorial Union law, the aspects described in paragraph 1 may be part of the quality management systems pursuant to that law."

Interpretation: The quality management system can be used to comply with the requirements of the first paragraph of the AI Act.

Financial Institutions

3. "For providers that are financial institutions subject to requirements regarding their internal governance, arrangements or processes under Union financial services legislation, the obligation to put in place a quality management system with the exception of paragraph 1, points (g), (h) and (i) shall be deemed to be fulfilled by complying with the rules on internal governance arrangements or processes pursuant to the relevant Union financial services legislation. In that context, any harmonised standards referred to in Article 40 of this Regulation shall be taken into account."

Interpretation: Banks and financial institutions must meet the governance requirements of the EU financial legislation. These institutions can meet their governance requirements by implementing a quality management system such as ISO 9001. However, there are a few restrictions that must be considered.

EU AI ACT RISK MANAGEMENT SYSTEM REQUIREMENTS (RMS)

EU AI Act also requires a risk management system. What is a risk management system? ISO 42001 and ISO 31000 are examples of international risk management systems.

Establish Risk Management System and Process

> 1. "A risk management system shall be established, implemented, documented and maintained in relation to high risk AI systems."

Interpretation: Developers of high risk AI systems have to develop, document and maintain a risk management system. The risk management system may be ISO 42001 compliant or a tailored ISO 31000 management system.

Risk Management Iterative Process

> 2. "The risk management system shall be understood as a continuous iterative process planned and run throughout the entire lifecycle of a high risk AI system, requiring regular systematic review and updating. It shall comprise the following steps:"

Interpretation: The risk management system is used throughout the lifecycle of the system and is continuously evaluated to ensure its capability.

> a. "identification and analysis of the known and the reasonably foreseeable risks that the high risk AI system

EU AI Act

can pose to the health, safety or fundamental rights when the high risk AI system is used in accordance with its intended purpose;"

Interpretation: Known and anticipated AI system risks to the health, safety, and human rights are identified and analyzed for their purpose of use.

b. "estimation and evaluation of the risks that may emerge when the high risk AI system is used in accordance with its intended purpose and under conditions of reasonably foreseeable misuse;"

Interpretation: The high risk AI system must be used for its intended purpose. Risks are identified and evaluated based on its intended use and those risks that may be anticipated from misuse are also identified and analyzed.

c. "evaluation of other possibly arising risks based on the analysis of data gathered from the post-market monitoring system referred to in Article 61;"

Interpretation: Article 61 of the Act specifies 'Post Market Monitoring by Providers and Post Market Monitoring Plan for high risk AI Systems. Once systems are deployed, additional risks may arise. These need to be identified, evaluated, treated, and monitored.

d. "adoption of appropriate and targeted risk management measures designed to address the risks identified pursuant to point a of this paragraph in accordance with the provisions of the following paragraphs."

Interpretation: Risks that have been identified from the AI systems need to be treated and managed appropriately. Risk treatment is similar to risk management.

Mitigating AI Risks

2 a. "The risks referred to in this paragraph shall concern only those which may be reasonably mitigated or eliminated through the development or design of the high risk AI system, or the provision of adequate technical information."

Interpretation: The identified risks from the AI systems need to be 'reasonably' mitigated or eliminated in the design and deployment of the system.

Minimizing AI Risk

3. "The risk management measures referred to in paragraph 2, point (d) shall give due consideration to the effects and possible interaction resulting from the combined application of the requirements set out in this Chapter 2, with a view to minimising risks more effectively while achieving an appropriate balance in implementing the measures to fulfil those requirements."

Interpretation: The risk management system should consider the interaction of causes and effects resulting from the AI system and if possible mitigate these risks.

Residual AI Risk

4. "The risk management measures referred to in paragraph 2, point (d) shall be such that relevant residual risk of each hazard as well as the overall residual risk of the high risk AI systems is judged to be acceptable."

Interpretation: Residual risks are the risks left after the application of risk - controls. The residual risks should be within the risk appetite of the developers.

"In identifying the most appropriate risk management measures, the following shall be ensured:

a. "elimination or reduction of identified risks and evaluated pursuant to paragraph 2 as far as technically feasible through adequate design and development of the high risk AI system;"

Interpretation: AI risks should be eliminated or reduced to an acceptable level throughout the lifecycle of the system.

b. "where appropriate, implementation of adequate mitigation and control measures addressing risks that cannot be eliminated."

Interpretation: If risks cannot be eliminated, they must be risk - controlled to an acceptable level.

c. "provision of the required information pursuant to Article 13, referred to in paragraph 2, point (b) of this Article, and, where appropriate, training to deployers."

Interpretation: The design and deployment of the AI system should be transparent and explainable.

d. "With a view to eliminating or reducing risks related to the use of the high risk AI system, due consideration shall be given to the technical knowledge, experience, education, training to be expected by the deployer and the presumable context in which the system is intended to be used."

Interpretation: Two items are critical in determining appropriate risk management: 1. Technical knowledge, experience, education and training of the AI designers and 2. Context and purpose of the AI system.

Testing AI Systems

5. "High risk AI systems shall be tested for the purposes of identifying the most appropriate and targeted risk management measures. Testing shall ensure that high risk AI systems perform for their intended purpose and they are in compliance with the requirements set out in this Chapter."

EU AI Act

Interpretation: High risk systems should be tested to determine the appropriate risk measures and controls. The purpose of the testing is to ensure compliance with their 'fitness for purpose' in terms of meeting requirements or complying with the appropriate laws.

Testing Procedures

6. "Testing procedures may include testing in real world conditions in accordance with Article 54 a."

Interpretation: Testing of high risk systems is conducted in conditions of how they are designed and deployed.

Testing Throughout AI Product Lifecycle

7. "The testing of the high risk AI systems shall be performed, as appropriate, at any point in time throughout the development process, and, in any event, prior to the placing on the market or the putting into service. Testing shall be made against prior defined metrics and probabilistic thresholds that are appropriate to the intended purpose of the high risk AI system."

Interpretation: The development and deployment of an AI system follows a lifecycle. Testing of the system should be made against defined performance and social metrics and even likelihood thresholds.

Fitness for Purpose

8. "When implementing the risk management system described in paragraphs 1 to 6, providers shall give consideration to whether in view of its intended purpose the high risk AI system is likely to adversely impact persons under the age of 18 and, as appropriate, other vulnerable groups of people."

Interpretation: Analysis needs to be done on the AI system risks if there is a possibility of the system negatively impacting minors or other vulnerable groups.

Additional Risk Requirements

9. "For providers of high risk AI systems that are subject to requirements regarding internal risk management processes under relevant sectorial Union law, the aspects described in paragraphs 1 to 8 may be part of or combined with the risk management procedures established pursuant to that law." [28]

Interpretation: Developers of high risk AI systems need to develop internal risk management processes.

FINAL THOUGHTS

- The EU AI Act is becoming the global AI statute.

EU AI Act

- More countries will adopt or copy sections of the EU AI Act for their national regulations.

- EU AI Act has specific Quality Management System requirements, like ISO 9001.

- The EU AI Act has specific Risk Management System requirements like ISO 42001 and parts of ISO 31000.

- Canada, Australia and many countries will probably adopt ISO 42001 as their 'what' requirements for AI.

ISO 42001
AI MANAGEMENT SYSTEM

ISO 42001 is an artificial intelligence management system (AIMS). A management system is a set of interrelated policies, procedures, processes, and resources a company uses to achieve its objectives. ISO 42001 is the AI equivalent of ISO 9001 for quality management system. This chapter introduces the topic of ISO management systems and specifically AIMS.

ISO MANAGEMENT SYSTEM STANDARDS

ISO management systems are designed to be applicable to many types and sizes of companies, regardless of their industry or sector. They are designed to be compatible with each other, so a company can apply multiple ISO management systems without duplicating efforts.

ISO Management Systems

The benefits of an ISO management system include improved risk - controls, consistency, safer products, and reduced costs.

ISO has developed several management systems standards including:

- **ISO 9001:2015 Quality Management Systems:**
    - Company maintains a documented quality management system (QMS).
    - Company identifies customer, user, and interested party quality requirements.
    - Company designs and develops apps and services that meet customer, user, and interested party quality requirements.
    - Company produces and delivers conforming apps and services.
- **ISO 14001:2015 Environmental Management Systems:**
    - Company maintains an environmental management system (EMS).
    - Company identifies and assesses its environmental aspects and impacts.
    - Company applies environmental objectives and targets.
    - Company monitors and measures its environmental performance.
    - Company takes corrective action to prevent pollution and other adverse environmental impacts.
- **ISO 45001:2018 Occupational Health and Safety Management Systems:**
    - Company maintains an occupational health and safety management system (OHSMS).

ISO 42001 - AI Management System

- o Company identifies and assesses its occupational health and safety risks.
- o Company applies occupational health and safety objectives and targets.
- o Company monitors and measures its occupational health and safety performance.
- o Company takes corrective action to prevent occupational accidents and illnesses.

- **ISO 27001:2013 Information Security Management Systems:**
  - o Company maintains an information security management system (ISMS).
  - o Company assesses its information security risks.
  - o Company applies information security risk - controls.
  - o Company monitors and measures its information security performance.
  - o Company takes corrective action to prevent information security breaches.

The Next ISO 9001

ISO believes ISO 42001 may become as influential as the 9001-quality management system:

> "ISO/IEC 42001 is a globally recognized standard that provides guidelines for the governance and management of AI technologies. It offers a systematic approach to

addressing the challenges associated with AI implementation in a recognized management system framework covering areas such as ethics, accountability, transparency and data privacy. Designed to oversee the various aspects of artificial intelligence, it provides an integrated approach to managing AI projects, from risk assessment to effective treatment of these risks."[29]

WHAT IS AN AIMS?

The purpose of an ISO management system is to bring the parts of an organization together so it can also comply with requirements.

AI is a social and technical system. The purpose of the AIMS is to protect social factors such as enhance trust and preserve rights. The purpose of the AIMS is also to protect technical factors such as to assure human safety.

What Is An AIMS?

As AI rapidly grows in different industries, it is crucial for companies to manage it systematically. ISO 42001 is a risk management framework for applying and managing AI systems. ISO 42001 is becoming more important as AI shapes all sectors.

An AIMS, as described in ISO 42001, is like a set of connected parts in a company. ISO calls this an integrated management system, which may integrate an AIMS with an environmental management system. This is genius. The management system brings the different parts of an organization together under one management systems umbrella.

ISO 42001 - AI Management System

AIMS Rules

ISO 42001 is among the first international standards for managing AI in the rapidly changing tech world. An AI system can be very broad. It is critical to define the scope of the AI system that can be evaluated or certified by a third party for certification purposes.

The management system creates rules, processes, goals, and steps to achieve those goals when it comes to developing, providing, or using AI systems responsibly. ISO 42001 lays out the rules and gives advice on how to set up, follow, document, comply, and improve an AIMS.

Another way to think about ISO 42001 is as a global AIMS that lays out the policies, procedures, and work instructions for creating, running, maintaining, and continually enhancing AI systems.

AIMS Elements

AIMS incorporates the following:

- Identify the AI systems used by the company.
- Define the scope of the AIMS, which may include the company's AI systems and apps.
- Identify the AI processes needed to manage the AI systems, such as AI processes for developing, testing, deploying, monitoring, and improving the AI systems.
- Develop and apply the AIMS, including the necessary documentation and training.

ISO 42001 Best Practices

ISO 42001 addresses challenges of AI like trust, ethics, explainability, and transparency for companies. ISO 42001 provides a structured approach to handle the risks and opportunities linked with AI, balancing risk management, innovation, and governance.

ISO 42001 promotes responsible and ethical AI practices, offering a framework for managing risks, complying with regulations, and building trust in AI applications. It aligns AI management with best practices, contributing to the success and sustainability of AI applications. To manage AI systems with ISO 42001, the emphasis is on integrating the AIMS within a company's existing structure and processes.

AI Challenges

A critical challenge of an AI system is that like humans they are prone to mistakes. Like humans, AI machines can make bad decisions and hallucinate resulting in low confidence and trust:

> "Just like humans, AI systems can make mistakes. For example, a self-driving car might mistake a white tractor-trailer truck crossing a highway for the sky. But to be trustworthy, AI needs to be able to recognize those mistakes before it is too late. Ideally, AI would be able to alert a human or some secondary system to take over when it is not confident in its decision making. This is a complicated technical task for people designing AI."[30]

ISO 42001 MANAGEMENT SYSTEM STRUCTURE

ISO standards have a common structure. Think of these as a common Table of Contents.

ISO 42001 Structure

The structure of ISO 42001 is:

- **Title Page:** The title page of ISO 42001 includes the title of the standard, the edition number, the ISO reference number, and the date of publication.

- **Foreword:** The foreword introduces ISO 42001 and includes information about the purpose of the standard.

- **Introduction:** The introduction provides background information on ISO 42001 including its scope, objectives, and intended audience. It also includes definitions of key terms and concepts used throughout the standard.

- **Scope:** The scope section defines the boundaries of ISO 42001 and specifies the range of topics or activities covered.

- **Normative References:** This section lists other standards or documents referenced within ISO 42001 that are essential for its implementation.

- **Terms and Definitions:** The terms and definitions section provides a list of key terms used throughout the standard along with their definitions.

- **Requirements:** The requirements section outlines the criteria and procedures that must be met to comply with the standard. This is the core of the standard and includes the following sections:

    o Context of the Organization.

    o Leadership.

    o Planning.

    o Support.

    o Operation.

    o Performance evaluation.

    o Improvement.

- **Annexes:** ISO 42001 Annexes are supplementary sections that provide additional information, examples, or technical details related to implementing risk - controls. ISO 42001 has 4 Annexes:

    o Annex A is a reference for risk - control objectives and controls.

    o Annex B provides guidance for AI controls.

    o Annex C provides additional guidance for AI related objectives and risks.

    o Annex D describes the use of the AIMS with other management system standards.

- **Bibliography:** The bibliography lists additional sources or references that may be helpful for further reading or research on the topic covered by the standard.

AI ASSURANCE AND ACCOUNTABILITY

AI audit is a systematic and independent process for gathering facts to determine the level of compliance with a specified reference. The purpose of an AI audit is to identify and assess risks and opportunities, and to provide recommendations for improvement.

Internal or external auditors conduct AI audits. Internal AI audits are conducted by employees of the company being audited, while independent third-party companies conduct external AI audits.

Types Of AI Audits

There are three main types of AI audits:

- **First-party AI audits:** the company conducts these audits to verify its management system is compliant with ISO 42001. AIMS companies can also self-certify to ISO 42001 requirements by conducting first party audits.

- **Second-party AI audits:** These AI audits are conducted by a customer, user, and interested party or other interested party to evaluate the company's management system.

- **Third-party AI audits:** These AI audits are conducted by an independent certification body to verify the company's management system is compliant with ISO 42001.

72   Trust Me: ISO 42001 AI Management System

AI audits can be conducted at any time but are often conducted on an annual or bi-annual cycle. The scope of an audit varies depending on the complexity of the company's management system.

Assuring/Auditing Control Objectives

ISO 42001 and ISO 27001 are largely designed around control objectives. These are objectives in the AIMS and ISMS that the standards specify. ISO 42001 control objectives comprise more than two-thirds of the standard and are found in the standard's annexes. ISO 42001 control objectives are covered in the second part of this book.

Most regulations such as the EU AI Act and the NIST AI RMF guidelines focus on safety and ensuring human rights are not breached. AI control objectives can be viewed from several perspectives:

- **Safety objectives:** These technical objectives include data quality, uncertainty estimation, algorithms, misuse, large language models, safety, cyber defense, explainability, and fair decision making.

- **Rights preserving objectives:** Human rights are often called rights preserving objectives. These include equity, fairness, explainability, trust, and responsibility.

- **Risk management objectives:** ISO 42001 focuses on risk - control objectives. The purpose of these objectives is to identify, analyze, and mitigate risks and harms from AI. Many control objectives are found in ISO 42001 Annexes.

AI Assurance Benefits

AI audits are an important part of the ISO management system process. They assure companies are compliant with ISO 42001 and ensure their management systems are effective.

If an audit identifies non-conformities, the company needs to take corrective action. Also throughout the AIMS implementation, the company reviews its management system to identify areas for improvement.

AI Audit Assessment Benefits

The benefits of conducting AI audits include:

- Improved compliance with ISO 42001.
- Reduced costs through increased efficiency and waste reduction.
- Improved environmental performance.
- Improved occupational health and safety.
- Stronger information security.
- Improved customer, user, and interested party satisfaction.
- Increased market share.
- Improved brand reputation.

ISO 42001 CERTIFICATION

More than two-thirds of ISO 42001 is a list of risk - controls to become ISO 42001 certified through the Conformity Assessment process. So, how does a company get certified?

How to Certify to ISO 42001?

There are different ways to do third-party certification to ISO 42001. Here are some steps on how a company complies with ISO 42001:

1. **Understand the ISO 42001 standard:** The first step is to understand the requirements of ISO 42001. Purchase the standard. This involves studying the standard's clauses, controls, and Annexes A and B, which outline the controls organizations implement based on their risk assessment.

2. **Conduct a gap analysis:** Assess the organization's current AI system controls and practices against the requirements of ISO 42001. Identify gaps or areas where improvements are needed to achieve compliance and meet objectives.

3. **Establish leadership support:** Get support from executive management to ensure commitment to implementing and maintaining the AIMS throughout its lifecycle. Leadership support is crucial for allocating resources, defining roles and responsibilities, and driving the AIMS implementation process.

4. **Define the scope:** Determine the scope of the AIMS, including the boundaries of the AIMS, specifically which AI systems are covered.

5. **Follow a lifecycle approach:** AI systems follow a lifecycle from conception, design, use and disposal. Identify a framework to follow for the lifecycle approach. The

ISO 42001 - AI Management System

framework can be ISO 42001, ISO 31000, NIST AI RMF, and other standards.

6. **Perform risk assessment:** Conduct a comprehensive risk assessment to identify and evaluate AI system risks. This involves assessing risks, threats, vulnerabilities, and potential impacts.

7. **Develop policies and procedures:** Develop AI policies, procedures, and guidelines that address the requirements of ISO 42001.

8. **Implement risk - controls:** Implement risk - controls and measures to mitigate identified risks and address the requirements of ISO 42001.

9. **Provide training and awareness:** Train employees on AI policies, procedures, and best practices.

10. **Monitor and measure performance:** Establish processes for monitoring, measuring, and evaluating the performance of the AIMS. This may involve conducting internal audits, performing periodic reviews, and collecting metrics to assess the effectiveness of controls and compliance with ISO 42001.

11. **Conduct management review:** Regularly review the performance of the AIMS with executive management. This involves assessing the effectiveness of controls, identifying areas for improvement, and making necessary adjustments to ensure continual improvement.

12. **Achieve Certification:** While ISO 42001 compliance does not require certification, some organizations may choose

certification by third party bodies to demonstrate conformity with the standard or may even self-certify.

13. **Maintain and Improve:** Continuously maintain and improve the AIMS to adapt to changes in the organization, technology, and security threats. A company complies with ISO 42001 and establish a robust AIMS to assure safe and rights preserving apps.

Importance Of AI Conformity Assessment

Conformity assessment is important for several reasons. It:

- Shows a company is committed to AI, environmental performance, occupational health and safety, or information security.

- Assures AI apps and services meet customer, user, and interested party requirements.

- Reduces the risks of accidents, injuries, and environmental damage.

- Protects information assets from unauthorized access, use, disclosure, disruption, modification, or destruction.

- Improves a company's reputation.

AI LIFECYCLE APPROACH

ISO 42001 emphasizes a lifecycle approach to managing and assuring AI systems. What does this mean?

ISO 42001 - AI Management System

AI Systems Lifecycle

ISO 42001 provides a framework for ensuring safe development, deployment and use throughout its life. Companies can use the following lifecycle approach to manage AI systems:

- **Concept and planning:** This phase defines the purpose, use cases, and objectives of the AI system. This includes identifying stakeholders, defining the architecture, and developing the AI development plan. Both social and technical elements are considered in AIMS planning.

- **Design and development:** This phase defines the design, coding, and development of the AI system. Social principles involved in risk based, problem solving and decision making are understood to ensure fairness. Risk – controls are developed to mitigate bias in training data, ensure data integrity, and provide explainability in the risk based, decision making process.

- **Validation and verification:** Before deployment, the AI systems goes through rigorous testing and validation, so it performs as intended and is fit for purpose.

- **Deployment and operation:** Once software is validated the AI app can be deployed.

- **Monitoring and maintenance.** It is important to monitor the AI throughout its life. The AI system is monitored throughout its lifecycle for compliance, fitness, safety, performance and stakeholder satisfaction. If there are performance or other challenging issues, these risks are identified and mitigated through root cause evaluation and corrective actions.

- **Continuous improvement:** Continuous improvement is a key element of all ISO standards including ISO 42001. Improvement may mean retraining the AI model, adjusting the algorithms, or improving input data.

- **Decommissioning:** ISO 42001 also focuses on the AI systems end of life. It is critical the AI system is retired responsibly. This may mean disposing of the software from platforms or deleting personal data.

ISO 42001 BENEFITS

ISO 42001 is the first AI management system standard produced by the International Organization for Standardization. ISO 42001 is a global groundbreaking AI standard that will impact ISO 9001 certification and ISO 31000 conformance to the standard. Why? The EU Act is a law. Compliance with the Act for high risk AI systems is required. This will increase the adoption of ISO 9001 and ISO 31000 by a factor of 2 or more.

ISO 42001 Advantages

ISO 42001 advantages for companies dealing with AI app development or services include:

- **Global recognition:** ISO 42001 is internationally recognized, which helps communication, collaboration, and trade between organizations across different countries and regions.

ISO 42001 - AI Management System

- **AI governance:** ISO 42001 aligns with legal and regulatory standards, offering a structured framework so companies can adhere to laws and regulations. This proactive approach prevents legal pitfalls and shows a company's commitment to meeting legal requirements in AI.

- **Structured framework:** Organizations design and deploy using a standardized AIMS approach.

- **Compliance:** ISO 42001 helps organizations ensure compliance with regulatory requirements and industry standards.

- **Customer focus:** Complying with ISO 42001 standards demonstrates a commitment to meeting customer expectations, regulatory requirements, and stakeholder expectations.

- **Risk management:** ISO 42001 includes requirements and guidance related to risk management, helping organizations identify, assess, and mitigate AI risks that could impact their operations, reputation, or objectives.

- **Practical guidance:** ISO 42001 provides practical guidance on management strategies. Companies identify and address potential risks of their AI applications.

- **Generate revenue:** AI is set to become a major economic driver in the future. AI revenue growth is projected to be exponential.

- **Resource optimization:** ISO 42001 promotes efficient use of resources including human resources, materials, energy, and finances.

- **Stakeholder confidence:** Adhering to and complying with ISO 42001 improves stakeholder and user confidence in an organization's ability to deliver AI apps and products that meet social, quality, safety, and environmental standards.

- **Responsible AI:** This standard assures ethical use of AI by providing guidelines and principles. Companies consider the societal impacts of their AI applications building trust and addressing ethical concerns.

- **Reputation management:** ISO 42001 improves trust in AI applications aiding companies in maintaining a positive reputation.

- **Identifying opportunities:** ISO 42001 encourages innovation within a structured framework providing a systematic approach to explore and apply AI technologies. Companies find opportunities for improvement in their AI applications.

- **Continual improvement:** ISO 42001 promotes continuous improvement by providing guidance on how to identify and address areas of AI risk.

FINAL THOUGHTS

- The purpose of an ISO management system is to bring parts of an organization together into an integrated system of processes and risk - controls.

- Any company, big or small, that works with AI apps or services can benefit from ISO 42001.

ISO 42001 - AI Management System

- IS 42001 applies to many industries and is useful for public sector agencies, companies, and nonprofits.

- ISO 42001 is a guide for companies looking to manage their AI systems efficiently, effectively, and economically.

- This is important because using AI, especially its machine learning aspect, comes with complexities and questions.

- AI can be low risk to high risk depending on its use cases, context, fitness for use, and risk impacts.

- In general, the higher the AI risk, the more risk - controls in the AIMS.

- Each of the following chapters cover a major clause in the AIMS.

- **Caveat:** The AI guidelines may change when you read this book. Please consult with an expert prior to implementing the guidance in this book.

ISO 42001: AI CONTEXT

Each of the following chapters describes a section in the standard. It is important to understand that ISO management system standards are technical and often difficult to decipher.

In this chapter, we cover AIMS Context. We have gone beyond a simple paraphrase of ISO 42001 to present a more comprehensive and nuanced explanation of the standard. This hopefully will help the reader to not only understand AIMS requirements but appreciate their value and implement them effectively.

UNDERSTANDING ORGANIZATIONAL CONTEXT

A company starting its AI journey needs to identify the outside and inside factors that affect its ability to achieve its goals with its AIMS. For example, the company considers the laws and regulations and the needs and expectations of its customers, users, and interested parties.

What Is Organizational Context?

Usually, AIMS context is broken down into 1. External context and 2. Internal context.

External context involves the following competitive factors:

- Laws and regulations that apply to AI.
- Government AI policies and guidelines.
- Potential benefits and risks of using AI.
- Ethical implications of using AI.
- Competitive landscape for AI apps and services.

Internal Context

Internal context involves the following organizational factors:

- Values, culture, and ethics.
- Governance structure.
- Risk management framework.
- AI risk tolerance.
- Risk based, problem solving and decision making processes.
- Resources and capabilities.
- Stakeholder needs and expectations.

Key Context Questions To Ask About AIMS?

A common thread to this book is that evaluating AI is all about risk. To understand risk, it is important to ask the right questions. Early in the AIMS and risk evaluation journey, the following are questions to consider:

- What is the intended use of the AI app?
- Is there a use case (s)?

ISO 42001: AI Context

- Is the AI app suitable for purpose and the use case?

- How representative is the training dataset to the operational context?

- How was the model trained?

- How do the AI app's characteristics align to the responsible dimensions of the use case and context?

- What are limitations of its functionality?"[31]

Answering these questions help understanding the context, interested parties, and scope of the AIMS.

AI Context Matters

AI context is important for these reasons:

- **Provide understanding**: Provides the background information necessary to understand the boundary conditions, stakeholders, assumptions, and use cases of AI.

- **AI Clarity**: Helps to clarify ambiguous or vague statements by executive management, developers, marketers and others visualizing the systems. These vague statements have to be operationalized into a use case, project plan, and requirements document. By providing additional details or background information, context ensures the right problem is solved and decision is made.

- **Risk based, decision making**: ISO 42001 should be implemented so it improves risk based, decision making processes. Decision makers can rely on contextual information to assess risks, evaluate options, and determine the best method to manage AI systems.

- **Risk based, problem solving**: In risk based, problem solving scenarios, context provides valuable clues about boundary conditions, risks, and constraints that help identify potential solutions.

- **Communication**: Effective communication among users and AIMS stakeholders helps ensure consistency and safety of high risk AI systems to meet business goals and regulatory requirements.

- **Cultural sensitivity**: AI systems are social and technical systems. Developers and engineers focus on the technical systems. Social scientists and impacted communities are concerned about the social impacts. Social context helps navigate cultural differences and avoid misunderstandings among the app's stakeholders and users.

AIMS STAKEHOLDERS

One of the first questions to consider are: Who are the users and more broadly who are the stakeholders of the AIMS. This is seemingly a simple question, but it helps to identify the scope, objectives, and framing of the AIMS. AI system stakeholders can be those designing the AI app or those impacted by the app. AIMS interested parties can be a broader or even a smaller group.

Remember, AI apps are both social and technical systems impacting many. In ISO 42001, the term 'interested parties' is often used to describe humans that create the AI app as well as the many parties impacted by the app.

Interested Parties And Users' Requirements

Once the company identifies AI system build or use case, the company has to understand the needs and expectations of the people who are impacted by their AI systems. This includes customers, users, interested parties, employees, shareholders, and the public.

Company understands the needs and expectations of users by conducting surveys and analyzing data. Once the company understands the needs and expectations of the social and technical stakeholders, it can develop its AI guard rails, accountability, and assurance systems.

AIMS Stakeholders

The AIMS ecosystem has many stakeholders:

- **AI providers:** AI providers develop and sell software and services. They provide consulting or application services.

- **AI users:** AI users are those who use the AI system. Their needs and expectations need to be understood when making design, development, and risk decisions.

- **AI interested parties:** AI interested parties can be one or any of these stakeholders. Interested parties can be ISO consultants, AI consultants, certification bodies, AI auditors, and shareholders.

- **AI producers:** AI producers develop systems for their own use or for sale to others. For example, an AI producer may develop a system to improve its customer, user, and interested party service operations or to develop new apps. It is crucial to remember the AI app is both a technical and social system.

- **AI customers:** AI customer, users, and interested parties use systems to improve their operations or to deliver apps and services to their customers, users, and interested parties. For example, a customer, user, and interested party may use an AI system to automate tasks, improve customer service, or develop new apps or services.

- **AI partners:** AI partners work with other companies to develop and deploy systems. For example, a partner may work with a producer to develop a new AI system or with a customer, user, and interested party to deploy an AI system.

- **AI trainers:** AI trainers are humans whose data is used to train and operate systems. AI subjects may be customers, users, interested parties, employees, or other humans whose data is collected and used by companies.

- **AI regulators:** Authorities are government agencies that regulate AI. Authorities may develop regulations for the development, use, and deployment of the systems.[32]

SCOPING THE AI SYSTEM

The scope or the boundary conditions of the AIMS determine what activities the company takes to architect, design, assure, and deploy the AIMS.

ISO 42001: AI Context

AIMS Scope

Companies need to decide and document what parts of their business is covered by their AIMS. This is called the scope of the AIMS. Scoping of the AI system is based on understanding the AIMS context and fitness of use and purpose.

The scope of the AIMS may depend on use cases, requirements, decision making capabilities, leadership, planning, support, operation, performance, evaluation, system improvement, risk - controls, and objectives.

AIMS Requirements

AI requirement is a statement specifying a necessary condition or action that is done to comply with ISO 42001. AI requirements can be expressed using:

- **'Shall' language:** This indicates the AI requirement is mandatory and is met without exception.

- **'Should' language:** This indicates the AI requirement is recommended but is not mandatory.

- **'May' language:** This indicates the AI requirement is optional and may not or cannot be adhered to.

Solving The Right Problem

One of the first questions that we ask in our consulting is: 'what problem are we trying to solve?' If an engineer, developer, consultant, auditor, or manager does not know the answer to this question, then a lot of time is spent going in the wrong direction ending up in dead ends.

This question comes down to understanding the problem to be solved or the decision to be made. It involves understanding context. So, context matters. Without AI systems context, it is easy to misinterpret or misunderstand the use case impacts of the AI system.

FINAL THOUGHTS

- This chapter and the subsequent chapters cover a section of ISO 42001.

- Context is the environment in which the AIMS operates.

- Context can be divided into external context and internal context.

- Company starting its AIMS journey needs to identify the internal and external factors that define the AIMS context so it can meet its safety requirements and meet its objectives.

- Context matters because it provides understanding of the purpose and fit of the AIMS; clarify its development; and facilitate decision making.

- AIMS stakeholders include providers, users, interested parties, regulators, designers, testers, and other stakeholders.

- Scoping of the AIMS means defining its context and boundary conditions.

ISO 42001: AI LEADERSHIP

In this chapter, we cover the importance of AI leadership, policies, roles, and responsibilities. AI leadership in ISO 42001 includes the Board of Directors and executive management involved in approving and resourcing the AIMS.

LEADERSHIP ENGAGEMENT

The governing body is the highest level of risk decision making authority within a company. This can be a company's board of directors, but it may be a council, a committee, or another group of humans. The role of the AI governing body is becoming increasingly important as companies face complex AI systems and demanding AIMS challenges.

Governing Body

The governing body is often the Board of Directors if it approves the AI app and AIMS. This leadership level is responsible for setting the company's strategic direction, allocating app resources, and ensuring the company complies with applicable laws and regulations. It has oversight of the company's management systems, including the AIMS, environmental management system, and occupational health and safety management system.

In some cases, the governing body can be executive management or the Program Management Group who authorizes the AIMS. By demonstrating leadership and commitment to the AIMS, executive management assures the company's systems are used in a safe, responsible, and ethical way.

Board Level Support
The governing body has a clear understanding of the company's business and its objectives. It is aware of the risks and opportunities facing the company's AI development. The governing body meets regularly to discuss the company's performance and to make risk decisions about its future direction.

Executive Management?
Executive management refers to the highest management level of decision making authority within a company. This includes the CEO, and other senior executives. Executive management is responsible for setting the company's strategic direction, allocating resources to design the systems, and assuring the company complies with applicable laws and regulations. This is becoming more important as more regulations are developed such as EU AI Act.

Executive Management Support
The EU AI Act and ISO 42001 emphasize the role of executive management in setting up and maintaining the AIMS. Executive management is required to:

- Demonstrate leadership and commitment to the AIMS.

- Assure the AIMS is integrated into the company's social, technical, and business processes.

- Communicate the importance of AI throughout the company.

ISO 42001: AI Leadership

- Review the AIMS on a regular basis.
- Assure the AI system is safe and rights preserving.
- Create AI policies and goals mapped with the company's plan.
- Make sure AIMS requirements are included in the company's business processes.
- Ensure the AIMS has the resources it needs.
- Stress the importance to manage and to follow AIMS requirements.
- Ensure AIMS achieves its goals.
- Assure the effectiveness of the AIMS.

AI POLICIES

Executive management authorizes AI policies. The company's top leaders create a plan for how AI use cases are aligned with the company's goals and assure applicable laws and rules are followed. These AI policies should be mapped to AIMS requirements.

ISO 42001 Requirements

ISO 42001 has 'shall' requirements for policy development. For example, executive management is required to develop AI policies, risk management framework, and objectives.

ISO 42001 requires AI policies, procedures, and documentation are also easy to access and apply. AI policies are also consistent with other policies in the company. Everyone in the company knows about the policies and people outside the company have access to it if they need them.

AIMS Policy Requirements

AIMS policies are:

- Available as documentation.
- Communicated to everyone in the company.
- Available to people outside the company.
- Define the governance, risk, and compliance roles of the AI governing body.
- Ensure the development and application of AIMS policies.
- Address the planning and management of AI risks.
- Cover the monitoring of risks throughout the AI app lifecycle.
- Improve the performance and reliability of the AIMS.
- Assure the ethical and responsible use of AI.

AI ROLES AND RESPONSIBILITIES

By taking these steps, executive management sends a clear message to the company that AI is a top priority by assigning roles, responsibilities, and authorities for the architecture, design, deployment, use, and retirement of the AIMS. If the system is not performing based on technical requirements, causes harm, or requires improvement or modification, then executive management is informed and ensures correction.

ISO 42001: AI Leadership

By assigning roles and responsibilities, executive management assures stakeholders the company is producing AI apps and services that meet the needs of its customers, users, and interested parties.

Management AI Authorizes

Another important element of leadership is to define AIMS authorities and responsibilities. The company's top leaders make sure everyone in the company knows who is responsible for what and when it comes to AIMS development and deployment. This includes things like who is responsible for developing AI systems, who is responsible for using systems, and who is responsible for monitoring the systems.

The top executives delegate the responsibility and authority to two key tasks: First, they make sure the AIMS complies with AI guidelines. Second, they report on how well the AIMS is performing to the Board.

Demonstrating Management Commitment

Executive management shows its commitment to the AIMS:

- Allocating time and resources to AI improvement initiatives.
- Setting clear goals and objectives.
- Recognizing and rewarding employees for their contributions to AI.
- Personally participating in AIMS audits and reviews.

FINAL THOUGHTS

- Leadership engagement is the essential requirement for AIMS success.

- Leadership engagement includes the Board of Directors, Executive Management, or Program Management.
- AI policies are the strategic plan for architecting, designing, deploying, and assuring the AIMS.
- AIMS policies flow down into procedures and work instructions.
- ISO 42001 requires roles, responsibilities, and accountabilities are clearly defined.

ISO 42001: AI PLANNING

In this chapter, we discuss AI planning, objectives, and change management. ISO 42001 planning defines the roadmap for the AIMS. When preparing for the AIMS, the company identifies the potential AI risks and opportunities, which are described in this chapter.

ACTION TO ADDRESS AI RISKS AND OPPORTUNITIES

Most of us think of risk in terms of consequences. ISO defines risk in terms of an upside opportunity and downside consequence.

Planning To Address AI Risks And Opportunities
In ISO 42001, risk opportunity assures that quality, safety, efficiency and interoperability of apps, services, and systems are safe and equitable.

Another way to look at opportunities in ISO 42001 is in terms of the potential advantages or benefits companies gain by adhering to and complying with these standards. For example, deploying ISO 42001 improves the quality of apps or services, improves organizational efficiency, increases customer satisfaction, and helps international trade.

AI Assurance

The concept of AI assurance is a common thread throughout the standard. AI assurance involves the following:

- AIMS is tailored to the context of the AI system.
- AIMS can achieve its objectives.
- AIMS establishes AI risk criteria.
- AIMS manages AI risk.
- AIMS is planned.

Purpose of AI Planning

AI planning is a critical element of ISO 42001. It is one of the sections of the standard that often needs the most interpretation. The organization needs to set up a plan with clear guidelines for understanding and evaluating AI risks.

These planning guidelines allow the organization to:

- Identify critical AI risks and opportunities. This involves identifying the critical few upside and downside AI risks and not the insignificant many.
- Carefully evaluate the significant risks involved with the AI system.
- Develop plans to address the identified risks.

ISO 42001: AI Planning

- Figure out how badly each risk could impact the organization.

Establish AI Risk Criteria

The concept of 'risk criteria' is a 'shall' requirement in ISO 42001. However, the standard really does not define how to establish and maintain 'risk criteria.' This concept is derived from ISO 31000 - 2018, the international risk management standard.

Establish Risk Criteria

Risk criteria are the standards, benchmarks, requirements, or objectives against which risks are evaluated and assessed. Risk criteria may include:

- **Framework identification:** What risk management framework is being used to evaluate AI risks. The framework can be ISO 42001, ISO 27001, ISO 31000, or AI RMF. The framework is important to level set the risk identification and evaluation process.

- **Risk identification and evaluation:** Risk criteria help the company evaluate the significant risks or in other words separate the significant few risks from the insignificant many. AI risks are compared against an objective or requirement and then prioritized based on their consequence and likelihood.

- **Benchmarking against an objective:** Risks in ISO 42001 are obstacles to reaching an objective. Risks can also be events and threats that stop a company from reaching an objective.

- **Establishing criteria benchmarks:** Risk criteria can serve as a standard or process for approaching AI risk management. Benchmarking can also be the basis for how to treat and manage risks, such as accepting, mitigating, transferring or avoiding AI risks.

- **Linked to strategic or operational objectives:** AI risks criteria should be aligned with strategic and operational objectives. Risk criteria should align with the organization's culture, goals, and tolerance for risk.

- **Evidence based, problem solving and decision making:** Risk criteria provide a standardized framework for AI risk based, problem solving and risk based, decision making.

AI Risk Assessment

Different industries and sectors have their own ways of understanding and assessing AI risks. It is the company's responsibility, during the AI risk assessment process to adopt a risk perspective that fits its context. This may involve using the definitions, taxonomy, framework, and processes of AI risks commonly used in the sector for which the AI system is being developed for and used.

Develop Risk Assessment Process

ISO 42001 does not prescribe a risk assessment process. It simply says a company must adopt one. The company needs to set up and develop a process for assessing AI risks that:

- Pulls information from the company's strategic goals and AI policies.

ISO 42001: AI Planning

- Assures repeated AI risks assessments result in valid and comparable outcomes.

- Identifies risks that either support or hinder the achievement of its AI objectives.

- Evaluates the AI risk likelihood and consequences for the company and humans and society if the identified AI risks occur.

- Determines the level and velocity of AI risks.

AI Risk Treatment

Risk treatment is the same as risk mitigation or risk management. There are 4 ways for treating risks: 1. Accepting, 2. Mitigating or controlling, 3. Transferring or 4. Avoiding AI risks. Risk treatment like risk assessment is not specified in ISO 42001. A company must develop its own risk treatment framework or rely on a standard such as ISO 31000.

Another option is to use the risk - controls found in the ISO 42001 Annexes. The second half part of this book lists 100's of controls for managing AI risks. Many of these controls are from ISO 42001 Annex A along with many additional controls.

Justifying AI Risk - Controls
One notable control is the requirement for justifying AI system development, outlining when and why the system is used along with the metrics to measure performance.

Documentation is also very important and is woven throughout ISO 42001. Documentation of design choices, machine learning methods, and evaluation of the AI system, and incorporating AI-specific measures of the AIMS.

The company keeps records of the steps it takes to identify and deal with risks and opportunities in AI. Instructions are also developed for applying AI risk management in companies that offer or use AI apps and systems.

Statement of Appropriate AI Controls

The design, type, application, form, extent, and choice of AI control is critical. ISO 42001 states a 'statement of applicability' about the choice of appropriate AI controls should be developed. This statement justifies the use and even the nonuse of an AI risk - control. This is important because ISO 31000 requires a company to tailor the risk management standard to the organizational context and use case.

ISO 42001 requires that:

- AI controls are tailored to the context.
- Controls are aligned to the organizational culture.
- Controls are justified and documented.
- Publicly available for scrutiny and not constrained by a non-disclosure agreement.

ISO 42001 Impact Assessment

ISO 42001 requires the likelihood and consequences of AI architecture, design, deployment and use cases be risk assessed and documented regarding the AIMS context, risks and use case.

Why is this important? If the AI system is high risk or low risk, ISO 42001 requires the AI impact assessment on the technical and social elements of the system. The impact assessment becomes a part of overall risk assessment, which becomes critical in the architecture, design, and deployment of risk - controls.

AI As A Social and Technical System

As discussed, AI is a social and technical system. As a technical system, AI systems are built on algorithms, large language models, coding, hardware, and infrastructure. These can be measured, documented, understood, managed, assessed, and risk - controlled.

However, public facing, AI systems often have social components such as human users, social decision making, legal requirements, divergent stakeholders, and social contexts. These often are not known, not understood, diverse, cannot be easily assessed, and cannot be managed.

ACHIEVING AI OBJECTIVES

AI objectives specifically required in ISO 42001 may include AI, environmental AI performance, occupational health and safety, and AI information security.

AIMS Objectives

The company must define its AIMS objectives and what it wants to accomplish with its AI system. Strategic and operation objectives are based on the organizational context, environment, regulations, business model, purpose, return on investment (ROI), and other contextual factors.

Trust Me: ISO 42001 AI Management System

Once the company understands the context in which it operates, it can develop and tailor the AIMS to achieve its goals and objectives. This assures the company's AI systems are used responsibly and ethically and align with the company's goals and objectives.

Use in AI Planning.

AI objectives are used and integrated throughout the ISO 42001 standard. AI objectives are used in AI planning, support, operations, treatment, and improvement.

ISO 42001 requires the company to address the following key questions in developing AI objectives:

- **What** does the company need to do regarding building, deploying and assuring the AIMS?
- **How** many resources are required to address the 'what' question?
- **Who** has the responsibility, authority, and resources to address the 'what' question?
- **When** will the 'what' question be achieved and deployed?
- **How** will the 'what' question outcomes be measured and evaluated?

Examples of AI Objectives

An AI objective is a Specific, Measurable, Achievable, Relevant, and Time-bound (SMART) target. The company develops AI objectives to support the achievement of the company's goals. They

ISO 42001: AI Planning

are used to guide risk based, decision making and to track progress.

AI objectives can be key performance indicators (KPI's) or simple goals. The following are a few examples:

- AI customer objectives may be to reduce customer, user, and interested party complaints by 10% in the next year.

- Environmental AI objective may be to reduce greenhouse gas emissions by 20% in the next five years.

- AI information security objective may be to apply a new security system to protect customer, user, and interested party data.

AI Objective Requirements

ISO 42001 provides some guidance on how to develop and apply AI objectives. ISO 42001 requires AI objectives are:

- Mapped to AI policy.

- Involve stakeholders in the objectives setting process.

- Specific, measurable, achievable, relevant, and timely (SMART).

- Communicated to interested parties when developed.

- Monitored, updated and documented.

- Address social and technical factors of the AI system.

PLANNING FOR AI SYSTEM CHANGES

Change happens especially in building AI systems. ISO 42001 requires change management.

Identifying Important AI Information

The company decides what information is important for both internal and external communication in the AIMS including:

- What information needs to be conveyed to whom by what date.

- Importance of the information.

- Methods for communication.

- Action required regarding the information with a due date and responsibilities and authorities.

- Corrective or risk management if required.

Managing Documentation

AI documentation from external sources the company is essential for planning and running the AIMS throughout the app's lifecycle.

Access to AI information is also controlled. Access involves a choice about whether someone is allowed to view or modify the written records. When it comes to managing AI design and AIMS documentation, the company addresses the following:

- Distribution, accessibility, retrieval, and use of AI documentation.

ISO 42001: AI Planning

- Storage and protection of AI documentation including ensuring the information remains accurate and reliable.

- Managing changes to AI systems such as through version control.

- Deciding how long the AI records are retained.

- Deciding how information are disposed of when they are no longer needed.

Managing AIMS Changes

To manage change, the company's AIMS addresses the following:

- Written records are mandated by the AIMS.

- Amount of written information required for an AIMS varies between companies.

- Company's size and the types of its activities, AI processes, apps, and services.

- Complexity of the AI processes and how they interact.

- Competence of the humans involved with AI and AIMS development.

- Detail the review and approval process to assure AI is suitable, safe, and adequate.

FINAL THOUGHTS

- ISO 42001 requires the AIMS achieves its AI objectives, minimizes or avoids unwanted risk consequences, and improves AI management and application.

- AI risk has both an upside component of opportunities and a downside of consequences.

- AI planning is conducted throughout the AIMS lifecycle.

- Risk assessment involves identifying and analyzing AI risks.

- Risk treatment is the same as risk management or risk mitigation.

- Risk treatment has 4 methods: 1. Accepting, 2. Mitigating or controlling, 3. Transferring or 4. Avoiding AI risks.

- Changes in the AIMS are planned and documented.

ISO 42001: AI SUPPORT

AI systems development is about change. New technologies are developed. Unexpected outcomes occur. So when a company architects, designs, deploys and assures it AIMS, AI support is implemented in an organized and structured way.

AI RESOURCES

ISO 42001 requires the organization has the necessary resources to design, implement, and maintain the AIMS. This resource requirement may involve:

- **Design, implement, and maintain the AIMS:** This includes human resources and documented procedures.

- **Operational efficiently and effectively:** AIMS processes and procedures are sufficient to deploy and measure planning processes, which assure objectives are met.

- **Continuous improvement:** AIMS resources are adequate and sufficient to assure AI systems and processes are monitored and improved.

Human Resources

Meeting this AIMS requirement involves training, mentoring, or reassigning current staff, or even recruiting or contracting skilled humans throughout the AI lifecycle.

Trust Me: ISO 42001 AI Management System

AIMS is different than most ISO management systems. AI technology requires higher knowledge, skills, and abilities from humans. AI is changing rapidly. Many AI technologies are unexplainable. These things make it difficult to find suitable humans to design and deploy the AIMS.

So, it is critical a company implementing an AIMS consider the following:

- Identify users, developers, and interested parties of the AIMS.

- Identify the skills and knowledge required for humans who are responsible for tasks architecting and monitoring AI performance.

- Make sure these humans possess the needed skills through suitable education, training, or experience.

Designing The AIMS

As mentioned, AI product development and AIMS are inherently risky. Humans designing the AIMS need to understand the AIMS context and the AI app's context including:

- AI policies, procedures, and work instructions for the architecture, design, deployment, use, maintenance, improvement, and retirement of the AI systems.

- Leadership needs to understand and describe the AI use cases.

- Parties impacted by the AIMS are diverse.

ISO 42001: AI Support

- Unintended consequences of the AIMS and AI app are evaluated.

- Risk consequences of not adhering to the AIMS requirements.

AI COMPETENCE

ISO 42001 emphasizes the importance of ensuring employees have the necessary competence to architect, design, deploy, and assure the AIMS.

AI Human Knowledge, Skills, and Abilities

AI human knowledge, skills, and abilities are key to implementing ISO 42001 management system. This comes down to the competence of the engineers, auditors, and professionals involved in building the AIMS and AI products.

Human competence is the ability to apply knowledge and skills to achieve intended results. Competence is essential for companies to manage their operations, deliver high risk AI apps and services that achieve their AI objectives.

Assuring AI Employee Competence

Companies assure the competence of their employees by:

- Developing AI training and development programs.

- Providing opportunities for employees to gain AI experience.

- Assessing employee AI competence.

- Providing AI feedback and coaching.

AI Support Benefits

By investing in the competence and engagement of their employees, companies reap several benefits, including:

- **Improved AI:** Competent employees are likely to produce high quality AI apps and services.

- **Increased productivity:** Competent employees are likely to work efficiently.

- **Reduced AI risks:** Competent employees are less likely to make mistakes that could lead to AI accidents or other problems.

- **Improved customer, user, and interested party satisfaction:** Competent employees are more likely to provide excellent customer, and user service.

- **Improved employee morale:** Competent employees are more likely to be engaged and motivated in their work.

AI AWARENESS

AI systems hallucinate. What does this mean? AI inputs, processes, and output are not understood and may not be explainable.

ISO 42001 requires AIMS stakeholders and users are aware of the use case, risks of noncompliance, and organizational AI policies.

AI Awareness Factors

AIMS is the ability to understand the AI system. This is often difficult if not impossible with GPT systems. They hallucinate. They are sometimes unexplainable.

AI decision making for critical infrastructure needs to be reliable and accurate such as for nuclear power plants and water plants. Public facing policy decisions need to be transparent and fair.

Factors to consider in AI awareness among stakeholders include:

- **Technical implications:** AI systems have inputs, assumptions, boundary conditions, context, criteria, stakeholders, and impacts.

- **Societal implications:** Public facing, AI decision making system can impact humans. These decision making systems may be biased, faulty, and simply wrong.

- **Understanding data:** AI systems are training on large language models that may be corrupt and biased.

- **Boundary conditions:** AI systems should be configured and designed so the boundaries of knowledge, data, and application are known.

AI COMMUNICATIONS

EU AI Act requires AI policies. AI policies are a formal statement of a company's intentions, principles, or AI guidelines. AI policies are set by a company's AI governing body or management. They are the key AI governance documentation. They are used to guide risk based, decision making and to assure the company's activities are consistent with its AI strategic objectives.

Software Documentation

Software documentation is critical for due care, due diligence, and compliance with the AIMS:

> "Organisations that develop software need to be able to demonstrate due diligence in the creation of that software, including using appropriate standards for logging: what code is written (as well as when, why, and by whom), which software and data libraries are used, and what hardware is used during system development.
>
> Organisations need to undertake and document appropriate testing before the software's release and perform monitoring and maintenance while the code is in use. Like other sectors, they should be held liable unless they can prove such due diligence. These procedures lead to a safer and more stable software systems, intelligent or not."[33]

ISO 42001 Documentation Guidance

ISO 42001 provides guidance on how to develop and apply AI policies. This guidance includes recommendations for:

- Identifying the need for AI policies.
- Consulting with stakeholders.
- Developing clear and concise language.
- Communicating the AI policies to employees.

ISO 42001: AI Support

- Reviewing and updating the AI policies.

AI Policies

AI policies are used to address topics, including AI environmental performance, occupational health and safety, and information security. For example, AI policies may state the company is committed to providing AI apps and services that meet customer, user, and interested party needs.

Why Are AI Policies Critical?

AI policies allow companies to:

- Achieve their AI objectives.
- Make better risk-based decisions.
- Assure compliance with laws and regulations.
- Reduce AI risks.
- Improve stakeholder communication and collaboration.
- Improve customer, user, and interested party satisfaction.
- Increase market share.
- Improve brand reputation.

AI DOCUMENTATION

Documentation has been a key component of ISO management system standards. For example, documented AI information is a key requirement of assuring that AIMS is risk - controlled and maintained by the company.

General AI Documentation

ISO 42001 documentation comes in many forms and can be from different sources. Documented AI information is used to support the company's operations and to provide evidence of compliance to ISO 42001.

Importance Of AI Documentation

Written records prove that AI processes are implemented and comply with the AIMS. Documented AI information can be:

- AI policies.
- AI design procedures.
- Work instructions.
- Flow charts.
- AI process maps.
- Use cases.
- AI tests.
- Testing data.
- Records.
- Reports.

Creating AI Documentation

The purpose of AI documentation is to:

- Provide a clear and concise record of the company's AI policies, procedures, and processes.

ISO 42001: AI Support

- Assure everyone in the company is aware of the company's AIMS requirements.

- Provide evidence of compliance with ISO 42001.

- Help communication and collaboration within the company.

- Support the company's continuous AI improvement efforts.

Controlled AI Documentation

ISO 42001 requires AIMS documentation used throughout its lifecycle is controlled. Controlled AI documentation means documents related to the AIMS has management practices and requirements such as 'shall' requirements. These requirements assure the documentation's accuracy, completeness, accessibility, and security. The company manages intended changes and assesses the AI outcomes of unintended changes and minimizing negative impacts.

Controlled Documentation
Controlled AI documentation may involve:

- **Access control:** Authorizing and authenticating access to proprietary and confidential information to approved users.

- **Version control:** Tracking changes to the documents over the AIMS lifecycle.

- **Review and approval of changes:** If changes are made to the AIMS or app, then these are recorded and approved.

Reasons Why AI Documents Are Controlled
There are several reasons why AI documentation is controlled:

- **Governance and compliance:** Documentation assures governance, management, and risk management compliance and provenance of the AIMS.

- **AI systems are complex:** Documentation provide a baseline of the AI design, assumptions, boundary conditions, use cases, stakeholders, requirements, use, maintenance, support, and retirement.

- **Improved AI:** Documentation assures apps and services meet customer, user, and interested party AI requirements.

- **Reduced AI risks:** Documentation identifies and addresses potential problems before they occur.

- **Auditability:** ISMS may be audited internally, by customers, or a Certification Body. The documentation provides transparency and audit trail.

- **Lessons learned:** Documentation provides a lesson learned for architecting, designing, deploying and assuring the AIMS.

- **Risks based thinking and decision making:** Documentation provides an audit trail for risk based, problem solving and decision making for building the AI system.

FINAL THOUGHTS

- ISO 42001 AIMS requires that the organization has the necessary resources to design, implement, and maintain the AIMS.

- AI systems are complex and often unexplainable.

ISO 42001: AI Support

- Adequate and sufficient resources need to be allocated to design, implement and maintain the AIMS.

- ISO 42001 emphasizes the importance of ensuring employees have the necessary competence to architect, design, deploy, and assure the AIMS.

- Public facing AI systems impact humans. Stakeholders and users of the AI system need to be aware of the possible use and impacts of the system.

- ISO 42001 requires the AI system is documented throughout its lifecycle.

ISO 42001: AI OPERATIONS

ISO 42001 requires companies to plan, implement, control their AIMS processes and operations within the scope of the AIMS. This chapter covers AI operations in the AIMS.

OPERATIONAL PLANNING/CONTROL

ISO 42001 operational planning involves defining the processes to achieve the desired requirements ('shalls') and objectives of the standard.

What Is Operational Planning?

AI operations are a set of interrelated or interacting processes and tasks that transform inputs into the AI product and app. AI operations are often very complex. They must be carefully structured and organized to achieve a business or process outcome.

Operational planning includes identifying the resources required such as people, equipment, and materials to implement the AIMS. For complex and high risk AI systems, operational planning means establishing the sequence of activities for critical AIMS processes throughout the lifecycle of the system. Planning also involves defining the risk criteria for an assessment.

Company Operations Regarding AI

ISO 42001 includes a set of requirements that cover aspects of the company's operations. These requirements involve environmental performance, occupational health and safety, or information security.

By complying with ISO 42001, companies demonstrate they are committed to meeting customer, user, and interested party requirements, protecting the environment, providing a safe and healthy workplace, and protecting information assets.

ISO 42001 Purpose

The main purpose of ISO 42001 is to guide companies in responsibly using, developing, monitoring, and securing the AI app through the AIMS.

The purpose of the AI systems is often defined in a use case. The AIMS can then be designed and integrated into the use case. Once this is done, the company decides what roles they want to play in relation to these AI systems. These include being an AI provider, producer, customer, user, and interested party, partner, subject, or authority.

Implementation of AI Ops

Implementation or deployment of the AI system needs to be planned throughout the AI systems lifecycle. AI documentation must be written, deployed and followed. Resources are required to assure AI system is safe and meets its objectives. Finally, operational monitoring and assurance are required to close the AIMS Plan, Do, Check, and Act loop.

Control of AI Ops

Risk and process control assure processes are performing effectively, meeting the defined requirements, mitigating risks, and

ISO 42001: AI Operations

meeting objectives. If there are unexpected risks, deviations, non-conformances or unintended consequences, then the organization reviews the change, mitigates the effects, and assures processes are stable, capable, and improving.

Benefits of Operational Planning And Control
Operational planning, implementation, and control result in the following benefits:

- **Improved operational efficiency:** AI processes are streamlined to eliminate waste and reduce the time it takes to develop and test code.

- **Improved AI safety:** AI processes are standardized to assure AI apps and services meet customer, user, and interested party requirements.

- **Reduced AI risks:** AI processes are designed to identify and address potential problems before they occur.

- **Increased customer, user, and interested party satisfaction:** AI processes are designed to meet the expectations of customer, user, and interested parties.

- **Improved employee morale**: AI processes are designed to be clear, logical, concise, and easy to follow, which makes work easier and more enjoyable for employees.

AI RISK ASSESSMENT

ISO 42001 emphasizes the importance of AI process management since companies can improve their AI performance. The challenge for evaluating AI systems, processes, and artifacts is that they are complex and often unexplainable. They involve different resources, and activities. ISO 42001 also provides little guidance on how to identify, map, analyze, improve, and AI risk - controls through the operations part of the lifecycle.

AI Risk Assessment

The challenge in implementing the AIMS is that risk assessment is essential to make AI understandable. However, ISO 42001 is weak in describing how this is conducted especially when an assessment is a prerequisite for establishing trust in the powerful technology:

> "Our trust in technology relies on understanding how it works. We need to understand why AI makes the decisions it does."[34]

The company needs to conduct AI risk assessments at scheduled times throughout the app's lifecycle or when important changes are suggested or take place. The company can keep written records of the outcomes of the AI risk assessments. AI risks are then assessed, compared, and contrasted against risk criteria.

AI Risk Appetite And Tolerance

The company also needs to create and maintain a set of rules for deciding which AI risks are acceptable and which are not. This process involves identifying and assessing AI risks and figuring out how to deal and how much damage they could cause.

AI RISK TREATMENT

When it comes to operational planning and AI risk - control, the company plans, executes, and oversees the AI processes necessary to fulfill its objectives. This involves setting standards for the

ISO 42001: AI Operations

AI processes and executing AI risk - controls in line with those standards.

AI Risk Treatment Methods

There are several AI risk treatment methods:

- **Risk avoidance:** Eliminate risks by changing the AI system's design or functionality.

- **Risk mitigation:** Reduce the likelihood or consequence of the risk through various techniques like data cleansing, bias detection methods, or implementing safety controls.

- **Risk transfer:** Share the risk with other AI parties through insurance or partnerships.

- **Risk acceptance:** Decide to live with the risk if the benefits of the AI system outweigh the potential social harms, technical constraints, and safety risks.

AI Risk Treatment Plan

The company needs to develop and implement an AI risk treatment plan and confirm it is working as intended. If new AI risks are identified during the AI risk assessment, additional risk treatment is implemented for the newly identified risks.

In cases where the AI risks treatment options described in the AI risks treatment plan prove to be ineffective, additional controls and treatment options need to be reevaluated using the risk treatment plan. The company has written records of the results of the risk treatment.

AI BUSINESS IMPACT ASSESSMENT

The company conducts an AI impact assessments at specified times or when substantial changes in the AIMS are expected. Additionally, the company maintains written records of the outcomes of AI business impact assessments.

AI Impacts

AI systems have a big impact on people's lives. For example, an AI system could be used to make risk decisions about who gets a loan could affect someone's ability to buy a house.

Company determines the risks and opportunities of AI systems deployment. The company need to get input from the humans affected by systems when developing, applying, maintaining, and monitoring the AIMS.

FINAL THOUGHTS

- AI operations are a set of interrelated or interacting processes and tasks that transform inputs into AI system and app outputs.

- Risk assessment is a critical element of AI operations.

- Risk assessment as part of the AIMS is conducted throughout the AI lifecycle.

- Risk assessment is important to separate the critical few risks from the insignificant many.

- Each risk assessment may trigger an additional risk treatment.

ISO 42001: AI Operations

- AI systems can impact social and technical systems.

- AI business and human impact assessments should be conducted throughout the AI lifecycle.

ISO 42001: AI PERFORMANCE EVALUATION

Once the AIMS is established, the company maintains and continually improves it. This chapter covers how to keep the AIMS up-to-date and ensure it is meeting company objectives.

MONITORING, MEASUREMENT, AND EVALUATION

AI monitoring is an essential part of each ISO management system, since it assures the system is achieving its AI objectives.

Monitoring AI Systems

AI monitoring is the process of regularly assessing the performance of an AI process or system to identify deviations from the desired state.

Key monitoring questions to consider include:

- What is the process to audit and verify AI app performance?
- What are the app's performance metrics?
- How can end users interpret the output of the AI app?

- Is the app continuously monitored for failure and other risk conditions?

- What implicit social and technical biases are embedded in the technology?

- How are aspects of trustworthiness assessed and how frequently?

- Can the app be retrained for fairness, explainability, or trust?

- Is the app explainable?

- How will the AIMS and app be monitored?

- What are the safety controls to prevent this system from causing damage?

- How can these controls be tested?[35]

Measuring AI Systems

AI measurement is the process of figuring out if AIMS objectives and requirements are being met. If requirements are not being met, AIMS is not stable or capable or there are unacceptable deviations, then these problems are addressed.

Making accurate and reliable AI measurements is essential for ensuring apps and services meet specifications. ISO 42001 companies assure AI measurements are accurate and reliable.

AI Measurement
There are two main types of AI measurement:

ISO 42001: AI Performance Evaluation

- **Quantitative AI measurements:** These AI measurements result in a numerical value such as statistical reliability evaluation of the apps performance.

- **Qualitative AI measurements:** These AI measurements result in a non-numerical value such as using a heat map to determine AI governance and compliance.

Accurate Measurements

ISO 42001 provides guidance on how to make accurate and reliable AI measurements. These standards cover topics such as:

- Selecting the appropriate AI measurement method.
- Calibrating AI measurement instruments.
- Risk - controlling measurement uncertainty.
- Recording and reporting measurement results.

Types of AI Monitoring

AI monitoring may involve:

- **Process monitoring:** AI process monitoring focuses on the performance of AI processes. AI process monitoring identifies AI process bottlenecks, process errors, or process deviations.

- **AIMS system monitoring:** AI system monitoring focuses on the performance of the AIMS. System AI monitoring identifies system-wide trends, issues, and inefficiencies.

- **Data testing:** Data testing ensures training data to build the model is representative, diverse, and free from biases.

- **Model testing:** Model testing chunks the AI system into components to ensure it functions as required.

- **Black box testing:** Black box texting involves testing outputs based on different inputs and prompts.

- **Performance testing:** Performance testing checks the apps performance for state management, accuracy, latency, and resource usage.

AI Monitoring Guidance

By following ISO 42001, a company designs its monitoring programs to achieve its objectives and improve its AI performance. ISO 42001 provides guidance on how to apply AI monitoring programs.

- Selecting the appropriate AI monitoring parameters.
- Setting up AI monitoring thresholds.
- Collecting and analyzing AI monitoring data.
- Taking corrective action based on AI monitoring data.

Why Monitoring Is Important?

AI monitoring allows companies to:

- Improve the performance of their AI processes and systems.
- Identify and address risks and problems early on
- Prevent problems from occurring in the first place.
- Make better risk decisions.

ISO 42001: AI Performance Evaluation

- Demonstrate compliance with ISO 42001.

Safeguarding Personal Information

ISO 42001 emphasizes protecting PII but provides little guidance. Companies handling PII in their AI systems should apply appropriate safeguards to protect from unauthorized access, use, disclosure, alteration, or destruction. This prevents identity theft, fraud, and other harm.

AI systems may use personally identifiable information. This includes information used to identify an individual such as:

- Name.
- Address.
- Phone number.
- Email address.
- Social Security number.
- Driver's license number.
- Medical records.
- Financial records.

AI Systems Security

ISO 42001 security requirements can be found throughout the standard. AI information security refers to the protection of information assets from unauthorized access, use, disclosure, disruption, modification, or destruction. Information assets include information such as data, documented AI, software, and hardware.

AIMS Security Considerations

ISO 42001 is not an information security management system standard. However, AI systems often include cyber security considerations, which should be considered:

- Identifying information assets and their value.
- Assessing the AI risks to information assets.
- Applying appropriate security AI risk - controls.
- AI monitoring and reviewing information security.
- Responding to AI information security incidents.
- Restricting access to information assets to authorized personnel.
- Training employees on AI information security.
- Having a plan for responding to AI information security incidents.

INTERNAL AUDIT

A company with an AIMS can use internal audits to evaluate AI performance. The company assesses both the AI performance and the efficiency of the AIMS.

Internal Audits

The company develops, deploys, and executes one or more audit programs for AI technical and social compliance. These programs include how AI audits are conducted, the methodologies used, who is responsible, planning criteria, and reporting of the AIMS. While creating the internal AI audit program, the company considers the significance of the AI processes and findings from prior AI audits.

An internal AI audit usually addresses:

- What is observed and measured?

- What techniques are used for AI monitoring, measuring, analyzing, and evaluating AI outcomes.

- What are the aims, standards, and range for the AI audit.

- Who are the AI auditors?

- How are AI audits conducted so they assure a fair process.

- How to ensure findings from AI audits are communicated to the appropriate managers?

- When to conduct the AI monitoring and measurement?

- When to assess the results from AI monitoring and AI measurements.

MANAGEMENT REVIEW

Senior leadership periodically assesses the company's AIMS to confirm it remains appropriate, sufficient, efficient, and trustworthy. The goal is to ensure regular management reviews and audits will increase understanding and explainability of the AIMS. But there are challenges:

> "But AI systems have a significant limitation: Many of their inner workings are impenetrable, making them

fundamentally unexplainable and unpredictable. Furthermore, constructing AI systems that behave in ways that people expect is a significant challenge. If you fundamentally don't understand something as unpredictable as AI, how can you trust it?"[36]

Management Review

AI management review is an evaluation of AIMS performance to ensure requirements and objectives are met. AI performance is measured using safety, trust, explainability, financial measures, and operational measures.

ISO 42001 emphasizes the importance of measuring and managing AI performance, since companies identify areas for improvement, make better risk decisions, and achieve their AI objectives.

General Management Evaluation

The management evaluation may include:

- Progress of tasks from earlier AIMS assessments.

- Adjustments, fixes, corrective actions, and preventive actions from the AIMS.

- Consideration of shifts in context, use case, input data, monitoring/measure processes.

- Review of operational, process and project performance throughout the AI systems lifecycle.

- Review of AIMS performance including patterns in risks and corrective actions.

ISO 42001: AI Performance Evaluation

- Outcomes of AI monitoring and measurement.

- Results from AI audits and risk evaluations.

AI Performance Benefits
The benefits to measuring and managing AI performance include:

- **Improved risk decision making:** Companies identify areas for improvement, make better risk decisions about resource allocation, and track progress towards AI goals.

- **Improved accountability:** AI measurement assures employees are aware of their responsibilities and are accountable for their AI performance.

- **Increased motivation:** AI measurement provides feedback on the performance of AI systems.

- **Improved customer, user, and interested party satisfaction:** AI measurement identifies areas where customer, user, and interested party satisfaction is low. This provides valuable insights for improving customer, user, and interested party service.

Management Review Feedback
The AI outcomes of the management assessment can involve opportunities for improvement and modifications to the AIMS. Written records provide proof of the findings of management reviews.

FINAL THOUGHTS
- AI performance is measured, documented, and monitored through the AIMS lifecycle.

- Monitoring can include AI system and process monitoring.
- ISO 42001 requires various types of monitoring such as internal auditing and management reviews.
- AIMS audits can be internal, second party or third-party AI audits.

ISO 42001: AI SYSTEMS IMPROVEMENT

Continual improvement refers to identifying opportunities to improve the AIMS and correct deficiencies in the system. It is a fundamental principle of ISO management systems. This chapter covers the continual improvement of the AIMS, which is a theme woven throughout ISO 42001.

CONTINUAL IMPROVEMENT

Continual AI improvement is not a one-time event. It is an AI process embedded into the company's culture and operations. It requires commitment from everyone in the company to identify and apply improvements. It requires experimentation and learning from mistakes.

Benefits of Continual Improvement

There are many benefits to applying continual AI improvement. These benefits include:

- **Improved AI:** Continual improvement identifies and addresses AI issues resulting in apps and services that exceed customer expectations.

- **Increased efficiency:** Continual improvement streamlines AI processes and eliminates waste leading to greater efficiency and productivity.

- **Reduced costs:** Continual improvement identifies and eliminates unnecessary costs leading to improved profitability.

- **Improved customer, user, and interested party satisfaction:** Continual improvement identifies and addresses customer and user satisfaction.

AI Continual Improvement

ISO 42001 emphasizes the importance of both AI effectiveness and efficiency as they are both essential for company success. AI effectiveness is the extent to which planned activities are realized and results are achieved. It is a measure of how well a company is doing in terms of meeting its AIMS objectives and achieving its outcomes. AI effectiveness is contrasted with efficiency, which is a measure of how well a company is using its resources.

Achieving AI Effectiveness

Several factors contribute to a company's AIMS effectiveness.

- **Clear and concise AI objectives:** Company has clear and concise objectives that are aligned with its AI goals.

- **Planning:** Company has a plan for achieving its AI objectives. The plan must be realistic, achievable, and measurable.

- **Communication:** Company communicates its AI objectives and plans to stakeholders including employees.

Employees understand what is expected and how their work contributes to the company's AI goals.

- **Resource allocation:** Company allocates its resources to support its AI objectives.

- **AI monitoring and evaluation:** Company monitors its AI objectives and evaluates their effectiveness.

- **Continuous AI improvement:** Company is committed to continuous AI improvement, risk management, and achieving their objectives.

AI NONCORFORMANCES AND CORRECTIVE ACTION

AI nonconformity or sometimes called nonconformance occurs when the AIMS and its processes do not meet requirements, standards, regulations, or expectations. It implies there is a deviation from what is considered acceptable or compliant in AI development, deployment, or operation. Nonconformities can be errors, defects, malfunctions, or other issues that may affect the AI system's performance, safety, ethics, or legality.

What Are Nonconformities?

Identifying and addressing AI nonconformities is a crucial part of AI assurance and risk management. Corrective action is taken to rectify nonconformities and bring the AI system and processes back into compliance.

When Nonconformities Arise

When an AI conformity arises or when something does not conform to the standards, the company can:

- Address the AI nonconformity.

- Take steps to apply AI risk - controls and rectify it.

- Deal with any repercussions of the nonconformities.

- Examine whether there is a need to address the root cause of the AI nonconformity to prevent its recurrence or occurrence in other areas.

- Review the AI nonconformity.

- Identify the underlying causes through a root cause evaluation.

- Identify if there are nonconformities or the potential to nonconformities waiting to occur in the AI lifecycle.

- Implement any necessary measures.

- Assess corrective actions taken.

- Adjust the AIMS and the app as required with additional corrective action.

Corrective Actions

AI corrective action may consist of measures to address and fix issues, problems, or non-conformities within an AI system. Corrective actions aim to resolve identified shortcomings, prevent their recurrence, mitigate risks, and assure the AI system operates as intended.

ISO 42001: AI Systems Improvement

The actions taken to correct nonconformities are equal with the impact of the identified nonconformities. There are written records of the nonconformities and the subsequent measures applied.

These actions may involve making changes to the AI algorithms, data, or AI processes to fix issues, errors, or biases. Corrective action is an important part of AI risk management and risk assurance.

ISO 42001 CHALLENGES AND BENEFITS

ISO 42001 is a new management system and framework. It is very useful for relatively low risk AI systems and apps. However there are several challenges for adopting and deploying ISO 42001.

ISO 42001 Challenges

The following is a partial list of challenges for using ISO 42001:

- **New standard:** ISO 42001 is a new management system standard. Companies may face a steep learning curve and investment applying the standard, training people, and dedicating resources to its implementation.

- **Vague standard:** ISO 42001 is often vague on requirements in the 10-page core standard. AIMS documentation for architecting, designing, deploying and assuring needs to be developed. For example, the list of objective controls in the Annex are often generic guidelines.

- **Very complex systems:** AI systems and apps are very complex. They are hard to evaluate and monitor. They are often unexplainable. They rely on large amounts of input and test data that are difficult to obtain and decipher. Companies do not want to share source code or algorithms.

- **Fairness and bias:** Fairness and bias in the AI systems can be difficult to detect and address.

- **Evolving technology:** AI is evolving quickly. Humans are having a difficult time keeping up and being able to evaluate AI systems.

- **Qualified humans:** AI engineers and ISO 42001 experts are very expensive and difficult to obtain.

- **Implementation costs:** ISO systems and apps are technical. The cost of the architecture, design, and deployment of the AIMS can be prohibitive for small companies.

- **Governance and compliance:** AI regulations require governance, risk, and compliance. Compliance costs are often prohibitive. The AIMS does not address many AI governance guidelines.

- **AI system risks:** EU AI Act requires the development of risk management systems and quality management systems for high risk AI systems. Conformity assessment of high risk systems is also a requirement. It is difficult to find Notified Bodies with these capabilities throughout the world.

FINAL THOUGHTS

- Continual improvement is a fundamental principle in all ISO management systems.

ISO 42001: AI Systems Improvement

- Continual improvement benefits include increased efficiency and effectiveness, reduced costs, and improved user satisfaction.

- If there are nonconformances, deficiencies, and risks in the AIMS, then these are root cause corrected.

- Corrective action follows a defined process to fix symptoms and root causes.

ANNEX RISK - CONTROLS

ISO 42001 has several Annexes. These Annexes are more than two-thirds of the standard. These annexes provide additional, detailed guidance for applying risk - controls to achieve AIMS objectives.

What Are Risk – Controls?

AI control - objective is what the AIMS specifies as a requirement to achieve or how to control the capabilities of an AI system to achieve safe (technical) and rights preserving (social) outcomes. Risk - control is a set of tools to manage risk. Risk - control is based on identifying an AI risk - control objective such as those required in ISO 42001 AIMS.

The ISO 42001 AIMS risk – control process can be summarized as:

- **Identify AI control – objective:** AI control - objective is the goal set for an AI system to achieve optimal performance. AI control – objective can be accuracy, efficiency, or ethical considerations. These objectives guide the development, design, deployment, and evaluation of the AI system.

- **Identify AI risks:** Risks are identified as things that can go wrong in complying and maintaining the AIMS.

- **Prioritize AI risks:** These risks are assessed and prioritized. Risks are not created equal. Some risks are more likely to occur or happen than others. Some risks can cause more damage. Some risks can be a threat and disrupt the AIMS. Some risks can cause noncompliance with laws and regulations.

- **Evaluate AI risks:** Then, prioritize risks based on likelihood and consequence.

- **Focus on significant AI risks:** It is important to focus on the significant few risks not the insignificant many.

- **Select and deploy risk – controls:** Select and deploy the appropriate AI risk - controls that lower the likelihood and consequence of a risk occurring.

Additional Risk - Controls

It is important to note the AIMS guidance may not be adequate or suitable for every AI system. Some AI systems may result in higher risks or are applied in critical infrastructure or social applications or may not align with a company's AI risk - controls. In this part of the book, specific ISO 42001 objectives are covered, but additional AIMS risk – controls are added.

HOW TO USE RISK - CONTROL ANNEXES
Annex A: Reference Control Objectives and Controls

Annex A is a generic guide or glossary on how to manage the development of AI systems. This Annex is a list of AI risk – controls that are implemented at various stages of the AI lifecycle from governance to risk to compliance (GRC). The risk – controls provide the starting point for architecting, designing, and deploying the AIMS.

Annex B: Implementation Guidance For AI Controls Listed in Annex A

Annex B is a reference for identifying and applying AI risk - controls identified in Annex A for managing risks. The AI risk - controls and their guidance are mapped to the risk - controls listed in Annex A.

Annex C: Potential AI Related Organizational Objectives and Risk Sources

Annex C lists AI objectives and identifies risk sources that are considered when managing AI. Some AI objectives may include fairness, transparency, explainability and security. Each AI objective has risks that can impede an organization from reaching its objective. By understanding each critical objective and its risks, the company can design its AIMS to achieve its control objectives.

Annex D: Use of AI Management System Across Domains Or Sectors

Annex D acknowledges AI is a wide and diverse domain used by many sectors. Each sector based on its context, apps, requirements, risks, and use cases may have different assumptions, stakeholders, and requirements. ISO 42001 must be tailored and integrated with domain specific standards to assure an effective, efficient, economic, and safe application of the AIMS.

ANNEX RISK - CONTROL ARCHITECTURE

The ISO 42001 Annexes compose most of the standard. ISO 42001 refers to these Annexes as normative. This means they incorporate material from ISO 42001 as well as other ISO standards relevant to the AIMS and risk.

Risk - Control Architecture

The rest of the book is organized into the following:

- **Section Title:** Titles include 'Policies Related to AI' and 'Internal Organization.'

- **AI Control Objective:** This is the overall risk - control objective for the section.

- **Sub Control Objective:** This is a sub control section. The sub control is part of the overall AI Control Objective.

- **AI Risk - Control:** Each Sub Control Objective has an 'AI Risk - Control.' The Risk - Control describes how the Sub Control Objective can be reached or met.

- **Deployment Of AI Risk - Control:** Each Sub Control Objective has 'Deployment of AI Risk - Control.' This describes how to deploy or implement the AI Risk - Control.

ISO 42001 Annex Material

The following sections cover AIMS objectives and risk - controls:

- **Policies Related to AI.**
    - AI Control Objective.

Annex Risk - Controls

- o Sub Control Objective: AI Policy.
- o Sub Control Objective: Alignment With Other Organizational Policies.
- o Sub Control Objective: Review Of AI Policy.

- **Internal Organization.**
  - o AI Control Objective.
  - o Sub Control Objective: AI Roles and Responsibilities.
  - o Sub Control Objective: Reporting Of Concerns.

- **Resources For AI System.**
  - o AI Control Objective.
  - o Sub Control Objective: Resource Documentation.
  - o Sub Control Objective: Data Resources.
  - o Sub Control Objective: Tooling Resources.
  - o Sub Control Objective: System And Computing Resources.
  - o Sub Control Objective: Human Resources.

- **Assessing Impacts Of AI Systems.**
  - o AI Control Objective.
  - o Sub Control Objective: AI System Impact Assessment Process.
  - o Sub Control Objective: Documentation of AI System Impact Assessments.

- o Sub Control Objective: Assessing AI System Impact on Individuals.
- o Sub Control Objective: Assessing Societal Impacts of AI Systems.

- **AI System Lifecycle.**
  - o AI Control Objective.
  - o Sub Control Objective: Management Guidance For AI System Development.
  - o Sub Control Objective: Objectives For Responsible Development of AI System.
  - o Sub Control: Processes For Trustworthy AI System Design and Development.
  - o Sub Control Objective: AI System Requirements and Specifications.
  - o Sub Control Objective: Documentation of AI System Design and Development.
  - o Sub Control Objective: AI System Verification And Validation.
  - o Sub Control Objective: AI System Deployment.
  - o Sub Control Objective: AI System Operation And Monitoring.
  - o Sub Control Objective: AI System Technical Documentation.
  - o Sub Control Objective: AI System Recording Of Event Logs.

Annex Risk - Controls

- **Data For AI System.**
    - AI Control Objective.
    - Sub Control Objective: Data For Development and Enhancement of AI Systems.
    - Sub Control Objective: Acquisition Of Data.
    - Sub Control Objective: Quality Of Data For AI Systems.
    - Sub Control Objective: Data Provenance.
    - Sub Control Objective: Data Preparation.

- **Information For Interested Parties.**
    - AI Control Objective.
    - Sub Control Objective: System Documentation and Information For Users.
    - Sub Control Objective: Understandability And Accessibility of Provided Information.
    - Sub Control Objective: External Reporting.
    - Sub Control Objective: Communication Of Incidents.
    - Sub Control Objective: Information For Interested Parties.

- **Use Of AI Systems.**
    - AI Control Objective.
    - Sub Control Objective: Processes For Responsible Use of AI System.

- Sub Control Objective: Objectives For Responsible Use Of AI System.
- Sub Control Objective: Intended Use Of The AI System.

- **Third Party Inter-Relationships.**
  - AI Control Objective.
  - Sub Control Objective: Allocating Responsibilities.
  - Sub Control Objective: Suppliers.
  - Sub Control Objective: Customers.

POLICIES RELATED TO AI

AI CONTROL OBJECTIVE

Company creates AI policies for managing the AIMS including architecting, designing, deploying, and assuring AI systems. AI policies are developed to comply with requirements and assure AI objectives are met.

SUB CONTROL OBJECTIVE: AI POLICY

AI Risk - Control
Company leadership leads, supports, and assures that AI systems meet the company's business needs and legal responsibilities. AI policy controls focus on meeting organizational, governance, and societal requirements.

Deployment Of AI Risk - Control
Deployment of AI policy controls includes:

- Develop company's business and AI strategy.

- Define governance and risk management framework.

- Ensure company context, vision, values and culture are integrated into AI systems.

- Define AI risks within the company.

- Define level of risks of AI systems and apps.

- Outline legal obligations, which include contractual AI obligations.

- Evaluate type of human oversight and control of AI systems.

- Define AI impacts on stakeholders, users, and interested parties.

- Define guiding principles and best practices that apply to AI systems and apps.

SUB CONTROL OBJECTIVE: ALIGNMENT WITH OTHER ORGANIZATIONAL POLICIES

AI Risk - Control
Company identifies areas where AI policies and objectives may influence other policies within the company.

Deployment Of AI Risk - Control
Deployment of controls includes:

- Conduct an assessment to identify whether and where existing AI policies may influence others.

- Manage AI resources and assets.

- Define AI domains such as security, safety, and privacy.

- Integrate AI into company's risk based, problem solving and decision making methodologies.

- Integrate AIMS into other management systems.

- Integrate AI into innovation, business model, operating model, and development processes.

- Define AI requirements for system impact assessments.

- Identify processes for developing AI systems.

- Specify policies how AI systems are developed, acquired, operated, and used.

- Develop consistent AI policies.

SUB CONTROL OBJECTIVE: REVIEW OF AI POLICY

Risk - Control
Company develops guidelines to establish and review AI policies. AI policies undergo periodic reviews to verify they remain appropriate and sufficient.

Deployment Of AI Risk - Control
Deployment of policy controls includes:

- Review social and technical policies against AIMS requirements and objectives.

- Conduct societal impact assessment analyzing social, economic, and cultural implications of AI policies.

- Review stakeholder engagement and input into AI policies and AIMS.

- Conduct risk assessment of AIMS and develop policies to mitigate risks.

- Review AI policies to improve and correct deficiencies.

- Review AI process include evaluating opportunities for enhancing the company's AI policies.

- Review methods of handling AI systems in response to changes in business conditions and regulations.

INTERNAL ORGANIZATION

AI CONTROL OBJECTIVE
Company establishes a system for ensuring the responsible execution, functioning, and supervision of AI systems.

SUB CONTROL OBJECTIVE: AI ROLES AND RESPONSIBILITIES

AI Risk - Control
Company defines and assigns roles and responsibilities for complying with AI requirements.

Deployment Of AI Risk - Control
Deployment of controls includes:

- Consider designating roles and responsibilities when reviewing policies and objectives throughout the AI lifecycle.

- Define data governance and privacy roles and responsibilities.

- Design governance and risk management framework.

- Assure algorithmic roles and responsibilities are defined to assure transparency and accountability for different areas of the company.

- Map roles for each element of ISO 42001 including risk management, system impact assessments, asset and resource management, security, safety, privacy, and development.

- Define responsibilities to identify and evaluate risks throughout the AI lifecycle.

SUB CONTROL OBJECTIVE: REPORTING OF CONCERNS

AI Risk - Control

Company develops a procedure for its employees to raise concerns they have about the company's involvement with an AI system across its lifecycle.

Deployment Of AI Risk - Control

Deployment of reporting system controls includes:

- Review reporting concerns for confidentiality, anonymity, or both.

- Be accessible and actively encourage reporting by both employed and contracted humans.

- Develop corrective action program.

- Ensure reporting of concerns is effective and anonymous.

- Specify appropriate investigative and resolution authorities for humans.

- Set up mechanisms for timely reporting and escalation to management.

- Ensure protection against retaliation by using anonymous and confidential reporting.

- Integrate existing reporting systems into AIMS requirements.

RESOURCES FOR AI SYSTEMS

AI CONTROL OBJECTIVE

Company manages its resources, which include AIMS components and assets to gain understanding and manage AI risks and risk consequences.

SUB CONTROL OBJECTIVE: RESOURCE DOCUMENTATION

AI Risk - Control

Company recognizes and records the resources needed at throughout the AIMS lifecycle.

Deployment Of AI Risk - Control

Deployment of resource documentation controls includes:

- Include critical components of the AIMS.

- Maintain records of AIMS resources.

- Identify data resources, which includes data used at various stages of the AIMS lifecycle.

- Ensure AIMS tools and utilities are documented such as AI algorithms, models, or software tools.

- Review AIMS and computing resource documentation like the hardware used for AI model development, execution, and storage for data and tools.

- Review and ensure human resources to operate and maintain the AIMS.

SUB CONTROL OBJECTIVE: DATA RESOURCES

AI Risk - Control
Company compiles information about the data resources employed in the AIMS.

Deployment Of AI Risk - Control
Deployment of data resource controls includes:

- Identify data's origin or source documentation.

- Update data when it is modified or updated.

- Identify the data categories such as training, validation, test, and production data.

- Review data categories and risk criteria of AIMS.

- Develop process for labeling data.

- Identify fit for purpose requirements and for which the data is used.

- Develop data preparation procedures.

Annex: Risk - Controls 165

SUB CONTROL OBJECTIVE: TOOLING RESOURCES

AI Risk - Control
Company compiles information concerning the tools and utilities used in the AIMS.

Deployment Of AI Risk - Control
Deployment of tooling resource controls includes:

- Define types, methods and tools used in machine learning.

- Review data collection and annotation tools for data, images, text and video.

- Identify types of algorithms and machine learning models and required tools throughout the AI system lifecycle.

- Identify tools or AI processes for data preparation including cleaning, preprocessing, normalization, and missing values.

- Determine techniques for model optimization, evaluation, and testing.

- Define methods for model evaluation.

- Identify resources required for provisioning tools.

- Identify tools that help model development and version control.

- Select software used for the design, development, and deployment of the AIMS.

SUB CONTROL OBJECTIVE: SYSTEM AND COMPUTING RESOURCES

AI Risk - Control
Company compiles data of the AIMS and computing resources employed in the AI system.

Deployment Of AI Risk - Control
Deployment of computing controls includes:

- Review resource requirements of the AIMS.

- Locate system and computing resources, which may involve on - premises, cloud computing, or edge computing environments.

- Identify AI processing resources, which include elements like network and storage resources.

- Identify information including software and computing resources used in an AI system.

- Review hardware used to execute AI system workloads, which could pertain to environmental effects.

- Identify resources necessary throughout the AI systems lifecycle.

SUB CONTROL OBJECTIVE: HUMAN RESOURCES

AI Risk - Control
Company compiles information about the human resources involved in the development, deployment, and operation of the AI system.

Deployment Of AI Risk - Control
Deployment of human resource controls include:

- Identify demographic groups to train models.

- Review demographic datasets used to train machine learning models.

- Consider AI requirements for a varied range of abilities.

- Determine the essential roles required for the AI system.

- Ensure there are sufficient data scientists and AI engineers.

- Make sure there is sufficient governance and monitoring of AI systems.

- Assure AIMS is integrated throughout the AI lifecycle.

ASSESSING IMPACTS
OF AI SYSTEMS

AI CONTROL OBJECTIVE

Company evaluates the effects of AI systems on humans and societies impacted by the AI system over its lifecycle.

SUB CONTROL OBJECTIVE: AI SYSTEM IMPACT ASSESSMENT PROCESS

AI Risk - Control
Company evaluates the possible AI outcomes for humans and societies from the creation to use of the AI systems.

Deployment Of AI Risk - Control
Deployment of system impact controls includes:

- Review potentially influence of AI systems on humans, communities, and societies.

- Review impacts of the AI system on different communities and marginalized groups.

- Assure fitness of use and fitness of purpose of AI systems.

- Identify sources, events, and potential AI risk outcomes.

- Analyze risk consequences and the likelihood of different AI outcomes.

- Assess risk decisions in acceptance and setting up priorities.

- Apply measures for mitigating AI risks.

- Document AI and ensure reporting and communicating the findings.

- Figure out the responsible entities to conduct the AI system impact assessment.

SUB CONTROL OBJECTIVE: DOCUMENTATION OF AI SYSTEM IMPACT ASSESSMENTS

Company records the outcomes of AI system impact assessments and storing these results for a specified period.

AI Risk - Control
Records are valuable resources for identifying information needs to be shared with users and interested parties.

Deployment Of AI Risk - Control
Deployment of impact assessment controls includes:

- Update AI system impact assessments regularly and when there are material system changes.

- Ensure AIMS policies are followed.

- Conduct internal audits and management reviews of impact assessments.

- Retain assessments based on regulations and internal requirements.

- Adhere to the company's retention of AI policies.

- Document and report positive and negative risk consequences of the AI system on humans and communities.

SUB CONTROL OBJECTIVE: ASSESSING AI SYSTEM IMPACT ON INDIVIDUALS

AI Risk - Control
Company evaluates and records the possible effects of AI systems on humans or groups of humans throughout its lifecycle.

Deployment Of AI Risk - Control
Deployment of human impact controls includes:

- Document required protections for various groups.

- Develop criteria for assessing fairness, transparency and explainability.

- Develop a process or audit for assessing impacts.

- Ensure the safety and health implications of the AI system.

- Ensure the AI system controls protect personal identity information and financial information.

- Ensure if personal rights are breached, humans and regulatory authorities are notified if required.

- Ensure if a breach occurs, it is corrected and reported.

- Audit the AI system to ensure compliance and effectiveness.

- Conduct a user experience and satisfaction survey of the AI system.

SUB CONTROL OBJECTIVE: ASSESSING SOCIETAL IMPACTS OF AI SYSTEMS

AI Risk - Control
Company examines and records the potential societal effects of their AI systems throughout the systems' lifecycle.

Deployment Of AI Risk - Control
Deployment of societal impact controls includes:

- Determine societal effects of AI systems based on the company's context and type of the AI systems involved.

- Identify goals to assess for societal impacts.

- Consider environmental sustainability, economic implications, governmental factors, health and safety concerns, and the influence on norms, traditions, culture, and values when assessing societal impacts.

- Identify stakeholders impacted by the AI system.

- Develop evaluation metrics and criteria.

- Develop framework and assessment methods.

- Identify societal effects of the AI system based on context, fitness for purpose, and type of AI system.

- Implement process for evaluating societal impacts by the AI system.

AI SYSTEM LIFECYCLE

AI CONTROL OBJECTIVE
Company defines and records AI goals and deploys procedures for the ethical and careful design and creation of reliable AI systems throughout its lifecycle.

SUB CONTROL OBJECTIVE: MANAGEMENT GUIDANCE FOR AI SYSTEM DEVELOPMENT

AI Risk - Control
Company specifies and records AI goals to assure the ethical development of AI systems throughout the development cycle.

Deployment Of AI Risk - Control
Deployment of system development controls includes:

- Pinpoint AIMS objectives that impact the design and development of the AI system.

- Integrate AIMS objectives into the design and development AI processes.

- Identify AI requirements and guidelines to assure measures are incorporated throughout AI lifecycle.

- Identify and review security practices, including anticipating, detecting, and preventing security threats.

- Ensure AIMS requirements and objectives are being met.

- Evaluate AIMS risk and treatments.

- Monitor the AIMS throughout its lifecycle.

- If there are deviations or nonconformities, conduct a corrective action.

SUB CONTROL OBJECTIVE: OBJECTIVES FOR RESPONSIBLE DEVELOPMENT OF AI SYSTEM

AI Risk - Control
Company records the conditions and specifications for introducing new AI systems or making important improvements to existing ones.

Deployment Of AI Risk - Control
Deployment of system impact controls includes:

- Identify use cases and the reasons for creating the AIMS, AI app, and its objectives.

- Identify stakeholders and interested parties of the AI system.

- Identify AIMS objectives for responsible AI system development.

- Establish risk criteria for the AI system.

- Develop risk based, problem solving and decision making procedures for evaluating AI system risks.

- Develop quality management system and risk management system used in the AI system.

- Identify criteria or metrics for measuring the system's success.

- Extend AI requirements across the AI system lifecycle.

SUB OBJECTIVE: PROCESSES FOR TRUSTWORTHY AI SYSTEM DESIGN AND DEVELOPMENT

AI Risk - Control
Company records the procedures for creating and developing the AI system.

Deployment Of AI Risk - Control
Deployment of trustworthy design controls includes:

- Define the processes and stages throughout the AI system's lifecycle.

- Figure out the AI requirements and methods for testing.

- Set up human oversight AI requirements, including AI processes and tools, especially when the system impacts humans.

- Identify the stages at which AI system impact assessments are conducted.

- Set expectations and rules for training data, including approved data sources and labeling.

- Specify the expertise or training required for AI system developers.

- Define criteria for system release.

SUB CONTROL OBJECTIVE: AI SYSTEM REQUIREMENTS AND SPECIFICATIONS

AI Risk - Control

Company develops a system to define AI system requirements and specifications throughout the product lifecycle.

Deployment Of AI Risk - Control

Deployment of AI system requirement and specification controls includes:

- Develop business, hardware and software requirements for the AI system throughout the lifecycle.

- Ensure business requirements cover context, business value, use cases, objectives, tasks, and deliverables.

- Ensure hardware requirements cover type of processor, video cards, memory, source, and storage.

- Ensure software requirements cover system, training models, testing, approvals, documentation, and change control.

- Review if AI system requires specialized requirements in automated requirements analysis using NPL, implicit requirements detection, and requirements traceability.

- Develop user and stakeholder feedback system for requirements review.

- Assure user feedback is incorporated into AI system.

SUB CONTROL OBJECTIVE: DOCUMENTATION OF AI SYSTEM DESIGN AND DEVELOPMENT

AI Risk - Control
Company creates a record of the AI system's design and development, aligning it with AI goals, documented requirements, and specified criteria.

Deployment Of AI Risk - Control
Deployment of documentation controls includes:

- Choose a machine learning approach (e.g., supervised, or unsupervised.

- Select the learning algorithm and the type of machine learning model to be used.

- Figure out how the model was trained and ensuring data AI.

- Evaluate and refine AI models based on social and technical requirements.

- Consider hardware and software components.

- Address security threats throughout the AI system's lifecycle.

- Determine the AI system's interface and how outputs are presented.

- Plan for human interactions with the AI system.

SUB CONTROL OBJECTIVE: AI SYSTEM VERIFICATION AND VALIDATION

AI Risk - Control
Company records verification and validation procedures for AI systems and provides criteria for their application.

Deployment Of AI Risk - Control
Deployment of verification and validation controls includes:

- Develop AI test methodologies and tools.

- Select test data that accurately represents the intended domain of use.

- Determine AI requirements for release criteria.

- Evaluate reliability and safety of AI requirements, including acceptable error rates for AI system performance.

- Determine AI objectives for responsible AI system development and use.

- Identify operational factors like data, intended use, and acceptable ranges throughout the app's lifecycle.

- Identify specialized use cases that may require more operational factor definitions or lower error rates.

SUB CONTROL OBJECTIVE: AI SYSTEM DEPLOYMENT

AI Risk - Control
Company creates a deployment plan and confirms that necessary AI requirements are satisfied before putting the system into operation.

Deployment Of AI Risk - Control
Deployment of system deployment controls includes:

- Determine if AI systems may be developed in one environment and deployed in another.

- Determine differences when creating the deployment plan.

- Evaluate possible use cases of the AI system outside of its intended purpose and fit.

- Develop a risk based, problem solving and decision making process for AI systems.

- Consider augmentation, automation, or replace of tasks.

- Build AI models capable of being proactive and predictive.

- Monitor AI system to gain insights on learning capabilities, and adaptive learning.

- Define prerequisites are fulfilled before the system is released and deployed.

- Develop a deployment plan based on risk management.

- Consider potential different risk impacts on stakeholders.

SUB CONTROL OBJECTIVE: AI SYSTEM OPERATION AND MONITORING

Risk - Control
Company records the essential components required for the continuous operation of the AI system.

Deployment Of AI Risk - Control
Deployment of operation and monitoring controls includes:

- Review AIMS and AI performance throughout its lifecycle.

- Monitor AI systems for general errors and failures and if the system is functioning according use case.

- Improve AI performance through machine learning, where production data and output data is used to further train the model.

- Review AI system for explainability, trust, bias control, and other key metrics.

- If AI performance changes, conduct root cause analysis to determine why.

- Tweak and correct the AI system if there are system errors and failures.

- Update the AI system throughout its lifecycle as the system evolves.

SUB CONTROL OBJECTIVE: AI SYSTEM TECHNICAL DOCUMENTATION

AI Risk - Control
Company develops tailored technical documentation for stakeholders.

Deployment Of AI Risk - Control
Deployment of technical documentation controls includes:

- Provide overview of the AI system including providing insights into its intended purpose, fitness for use, etc.

- Provide instructions for using and deploying the system.

- Describe technical assumptions of the AI systems deployment and operation.

- Identify key factors for deployment including runtime environment, software and hardware capabilities, and underlying data assumptions.

- Identify technical constraints and limitations, including criteria like acceptable error rates, accuracy, reliability, and robustness.

- Identify features of AI monitoring and functionalities that grant users ability to influence the system's operation.

SUB CONTROL OBJECTIVE: AI SYSTEM RECORDING OF EVENT LOGS

AI Risk - Control
Company develops event logs throughout AI system's lifecycle.

Deployment Of AI Risk - Control
Deployment of system event log controls includes:

- Monitor the AI system's operation.

- Make sure AI systems are set up to automatically capture and store event logs.

- Assure event logs are developed to automatically capture and store event data.

- Track the AI system's functionality to assure it is working as planned.

- Detect situations where the AI system performs outside of its intended conditions.

- Review undesirable AI outcomes or affect stakeholders.

- Determine data to capture in the event logs.

- Adhere to the AI system's purpose of use.

DATA FOR AI SYSTEM

AI CONTROL OBJECTIVE

Company understands the significance and risk consequences of data within AI systems during their development, deployment, or use across their lifecycles.

SUB CONTROL OBJECTIVE: DATA FOR DEVELOPMENT AND ENHANCEMENT OF AI SYSTEMS

AI Risk - Control
Company sets up policies, procedures, instructions, and records about safe AI systems development and improving the AIMS.

Deployment Of AI Risk - Control
Deployment of development of AI system controls includes:

- Identify privacy and security issues arising from the use of sensitive data.

- Monitor and control security and safety threats arising from AI system development.

- Assure AI system transparency and explainability throughout its lifecycle.

- Ensure and explain how data is used to determine the output of an AI system.

- Select representative training data and compare it to the operational domain of use.

- Assure accuracy and integrity of the data.

- Provide information on AI system lifecycles and data management.

SUB CONTROL OBJECTIVE: ACQUISITION OF DATA

AI Risk - Control
Company records how acquisition and selection data is used in AI systems.

Deployment Of AI Risk - Control
Deployment of acquisition of data controls includes:

- Acquire various data categories from diverse sources contingent on the AI system's purpose and application.
- Determine types and quality of data required for the AI system.
- Identify data sources reviewing the quality and representativeness of the data.
- Review the characteristics of the data source.
- Determine the demographics and characteristics of data objects.
- Determine handling of data and if it is biased.
- Determine provenance of the data.
- Detail data acquisition and use following data categories.

SUB CONTROL OBJECTIVE: QUALITY OF DATA FOR AI SYSTEMS

AI Risk - Control
Company specifies criteria for data quality ensuring the data employed in the development and operation of the AI system align with AI requirements.

Deployment Of AI Risk - Control
Deployment of data quality controls includes:

- Define data requirements for suitability, accuracy, reliability, and quality.

- Assure data meets the needs of the use case and interested parties.

- Assure suitability for fit for purpose and for intended purpose.

- Assess the completeness of the data to ensure it has all the information for the AIMS.

- Check for errors, missing values, biases, incompleteness, and gaps in the data.

- Review data for bias and the impacts of bias on system AI performance and fairness.

- Adjust the model and data to improve AI performance and fairness to acceptable levels for the use case.

- Define the AI system for training, validation, and testing.

SUB CONTROL OBJECTIVE: DATA PROVENANCE

AI Risk - Control
Company documents AI processes for verifying and recording the provenance of data over its lifecycle.

Deployment Of AI Risk - Control
Deployment of data quality controls includes:

- Define and document AI procedures to authenticate the origin of data used by the AI system.

- Develop and define risk management and data governance framework.

- Inventory the AI systems data and classify the data.

- Develop a data quality management program.

- Classify data based on quality standards, data profiles, validation, verification, and quality metrics.

- Define access controls to the data.

- Review data for privacy, completeness, compliance, bias and other criteria.

- Assure data integration, consistency, and interoperability across different systems and apps.

SUB CONTROL OBJECTIVE: DATA PREPARATION

AI Risk -Control
Company records its AI requirements for data preparation, outlining the methods and strategies employed in the AI process.

Deployment Of AI Risk - Control
Deployment of data preparation controls includes:

- Ensure data is suitable for AI use case and fitness for purpose.

- Assure data is suitable for AI tasks.

- Analyze data distribution and statistical techniques.

- Assure machine learning algorithms do not have missing or inaccurate entries or non-normal distribution.

- Clean data used in AI system addressing issues like correcting entries and handling missing data.

- Standardize data formats, units, and representations to assure consistency.

- Transform data to prepare for modeling.

- Ensure consistency among data classes and reduce size of large data sets.

- Increase diversity and quality of training data to assure fitness of use and purpose.

INFORMATION FOR INTERESTED PARTIES

CONTROL OBJECTIVE

Company provides stakeholders with information to understand and evaluate AI risks considering positive and negative impacts.

SUB CONTROL OBJECTIVE: SYSTEM DOCUMENTATION AND INFORMATION FOR USERS

AI Risk - Control
Company identifies and furnishes the required information to users of the system.

Deployment Of AI Risk - Control
Deployment of system documentation controls includes:

- Identify users and their needs of the AI system.

- Notify users the AI system is making decisions that may have impacts.

- Develop social and technical guidelines for the AI system.

- Determine fitness of purpose of the system.

- Document user interaction with AI.

- Develop procedures for overriding the system and when to do so.

- Provide information about accuracy and AI performance.

- Detail the impact covering potential benefits, harms, and AI risks.

- Modify claims of the system's benefits and notify users.

SUB CONTROL OBJECTIVE: UNDERSTANDABILITY AND ACCESSIBILITY OF PROVIDED INFORMATION

AI Risk - Control

Company develops procedures to ensure Information presented to interested parties is clear and understandable.

Deployment Of AI Risk - Control

Deployment of explainability controls includes:

- Ensure AIMS and AI system users develop understandable and explainable systems.

- Identify, analyze, and treat the risks in the AI system.

- Understand how the system makes predictions.

- Decipher how model changes with variations in input features.

- Recognize the diverse needs of various users and adapt the information to their understanding.

- Ensure information is easy to locate and accessible to users.

- Generate natural language explanations how model works.

Annex: Risk - Controls

SUB CONTROL OBJECTIVE: EXTERNAL REPORTING

AI Risk - Control
Company incorporates functionalities that enable the reporting of negative effects caused by the system.

Deployment Of AI Risk - control
Deployment of external reporting controls includes:

- Identify users and interested party reporting requirements of the AI system.

- Define purpose, scope, context, target user, and objectives of the AI system.

- Develop processes and procedures to report on the AI system.

- Ensure compliance and regulatory requirements are understood and adhered to.

- Identify data sources to train and operate the AI system.

- Describe model development and training methods of the AI system.

- Ensure technical, social, privacy, and security issues of the AI system are identified.

- Report a breach or failure in the AI system to critical users and authorities.

- Correct any problem ASAP.

SUB CONTROL OBJECTIVE: COMMUNICATION OF INCIDENTS

AI Risk - Control
Company defines AI strategy for informing users about system incidents and nonconformances.

Deployment Of AI Risk - Control
Deployment of incident communication controls:

- Understand incident legal reporting obligations, contractual agreements, and regulatory stipulations.

- Categorize types of safety incidents requiring communication.

- Develop incident risk management mitigations based on risk context and profile of AI app.

- Develop incident communication and reporting processes.

- Ensure timelines and list of authorities for notification.

- Determine details to be reported about the incident considering purpose and intent of use.

- Continuously monitor and improve incident reporting system.

SUB CONTROL OBJECTIVE: INFORMATION FOR INTERESTED PARTIES

AI Risk - Control

Company identifies its responsibilities for disclosing information about the AI system to stakeholders.

Deployment Of AI Risk - Control

Deployment of information controls includes:

- Identify stakeholders who require AI systems information.

- Identify type, detail, and frequency of information required by stakeholders.

- Determine information required by regulators.

- Determine if training data sets, code, validation, testing and algorithmic choices or other records are required.

- Develop process or procedure for sharing information.

- Provide risk management information and treatment options.

- Solicit feedback from shared information.

USE OF AI SYSTEMS

AI CONTROL OBJECTIVE

Company is responsible to ensure AI usage is in alignment with its policies.

SUB CONTROL OBJECTIVE: PROCESSES FOR RESPONSIBLE USE OF AI

AI Risk - Control
Company defines AI processes for the responsible use of AI systems.

Deployment Of AI Risk - Control
Deployment of responsible process controls includes:

- Develop processes, policies, practices to assure AI system is designed and deployed ethically.

- Develop a governance framework that outlines guiding principles for developing and deploying AI systems.

- Conduct a social and ethics assessment of the new AI system to identify and mitigate bias risks.

- Develop solutions to mitigate and treat social risks in the AI system.

- Implement contextual control to mitigate biases including in algorithms, data, and risk-based decision processes.

- Identify underrepresented groups of potential biases in the system.

- Protect user privacy and data security of users.

- Ensure regulatory compliance and notification.

- Develop assurance enforcing ethical standards.

SUB CONTROL OBJECTIVE: OBJECTIVES FOR RESPONSIBLE USE OF AI SYSTEM

AI Risk - Control
Company is responsible for ensuring the AI system is aligned and fit for purpose.

Deployment Of AI Risk - Control
Deployment of responsible use controls includes:

- Develop clear objectives that guide the responsible and ethical deployment of the AI system.

- Assure AI system is applied to comply with AI statutory guidelines and documentation in ISO 42001.

- Assure human in the middle and other oversight mechanisms are effective.

- Ensure continuous monitoring of the AI system.

- Ensure the AI system treats all groups fairly and without bias.

- Ensure accountability of the AI system with clear lines of responsibility and authority.

- Protect user privacy and data confidentiality.

- Develop access systems to ensure security, safety, and integrity of data.

- Ensure compliance with regulations and laws.

Annex: Risk - Controls

SUB CONTROL OBJECTIVE: INTENDED USE OF THE AI SYSTEM

AI Risk - Control

Company defines the intended use and objectives of the AI including how it will be designed, developed, deployed, and assured.

Deployment Of AI Risk - Control

Deployment of intended use controls includes:

- Define intended use, fit for purpose, use case, objectives, context, safety, target users, expected outcomes, and ethical considerations of the AI system.

- Define access controls to the AI system to restrict access and illegal use.

- Establish data governance policies and procedures to determine collection of training data and usage.

- Verify and validate the model to assure it meets objectives and requirements.

- Monitor and audit the system throughout its lifecycle.

- Assure regulatory compliance with governance systems.

- Ensure noncompliance and breaches are monitored, reported, and corrected.

THIRD PARTY INTER-RELATIONSHIPS

AI CONTROL OBJECTIVE

Company defines AI goals to assure the ethical and responsible use of AI systems among critical stakeholders and third parties.

SUB CONTROL OBJECTIVE: ALLOCATING RESPONSIBILITIES

AI Risk - Control
Company assigns responsibilities and authorities for ensuring the responsible development of AI systems.

Deployment Of AI Risk - Control
Deployment of responsibility controls includes:

- Assign roles, duties, responsibilities and authorities for the development and deployment of AI systems.

- Identify AI system owners responsible for the overall governance, accountability and oversight of the AI system throughout the systems lifecycle.

- Define AI system's objectives, stakeholders, use cases, fitness for purpose and compliance goals.

- Identify data manager responsible for security, quality, and governance of data.

- Design risk management and data management framework.

- Deploy the frameworks with defined accountabilities.

- Assure AI systems comply with regulations.

- Develop safety and rights review boards.

- Retain compliance personnel to assure regulations are met.

SUB CONTROL OBJECTIVE: SUPPLIERS

AI Risk - Control
Company secures a commitment from its suppliers to adopt a responsible approach in developing AI systems.

Deployment Of AI Risk - Control
Deployment of supplier controls includes:

- Identify what outsourcing activities is conducted with AI development, deployment, and assurance.

- Identify third party suppliers involved in AI systems.

- Determine supplier selection and due diligence to select suppliers involved in AI development.

- Develop contractual agreements that outline supplied deliverables, terms, and conditions.

- Ensure data protection, confidentiality, and privacy of proprietary and sensitive information.

- Develop quality assurance, control, and testing processes for supplied deliverables.

- Ensure data and algorithm security of supplied products and services.

- Identify and mitigate third party risks regarding contracted deliverables.

SUB CONTROL OBJECTIVE: CUSTOMERS

AI Risk - Control
Company incorporates a responsible approach to AI system development that considers the expectations and requirements of its customers, users, and interested parties.

Deployment Of AI Risk - Control
Deployment of system impact controls includes:

- Explain to customers the boundaries of the AI system.

- Explain to customers the AI system is in use with these precautions and decision making capabilities.

- Inform customers what personal data is captured, how it is used, and how it is protected.

- Offer customers control over their personal data.

- Offer customers a human solution and escalation option if possible.

- If there is a breach in data, notify customers immediately and regulatory authorities if required.

- Determine and test if there is algorithmic bias.

- Determine level of control by customers dependent on AI application risks and purpose of use.

- Assure the AI system is secure.

FINAL THOUGHTS

The AI control annexes are two-thirds of ISO 42001 standard. Benefits and advantages to the AI control objective approach involve:

Benefits

- AI control objective approach is like ISO 27001 approach to risk mitigation.

- AI control objectives should be used with low risk AI applications and systems.

- AI control objectives approach describe desired outcomes and objectives to provide AI trust.

- AI control objectives are straightforward to implement in an AIMS.

- AI control objectives approach identify and mitigate AI risks.

- AI control objectives can be incorporated into ISO 42001 AIMS.

Challenges

Challenges of using the AI control objective approach include:

- AI control objective approach is often transactional and work for low risk AI systems.

- High AI systems require a systemic and lifecycle approach to risk management.

- AI control objectives are new. Few in the conformity assessment community including auditors and consultants know how to implement these controls.

- AIMS does not address risk based, problem solving and decision making.

- ISO is encouraging a unified and integrated approach with existing management systems.

- AIMS is sufficiently different from other management systems that it may be difficult to integrate.

- High risk AI systems require higher assurance than the AIMS.

- Social elements of the AI system are difficult to assure and audit and will require approaches beyond control objectives.

GLOSSARY

4P's: Trademarked business risk model representing: Proactive, Preventive, Predictive, Preemptive® actions focused on risk management.

AI acquisition: Acquiring an AI system, while assuring fairness and explainability.

AI deployment: Implementing for use an AI system.

AI development: Architecting, designing, coding, and deploying an AI system, while assuring fairness and explainability.

AI evaluation: Regularly evaluating the performance of the AI system through business impact analysis and making system improvements.

AI explainability: Understanding how the AI system works, makes decisions, solves problems and being able to explain the process.

AI identification and planning: Determining the goals and intended use of the AIMS, risks, controls, and impact throughout its lifecycle.

AI Management System (AIMS): Set of interrelated processes and risk - controls for responsibly designing, developing, acquiring, deploying, maintaining, and retiring AI systems and services.

AI system impact assessment: Process often documented to analyze the impacts of AI on different users and businesses.

Accidental hazard: Source of harm or hazard created by error, negligence, or unintentional failure.

Accountability: Humans such as stakeholders, management and employees who are responsible for the AI systems decisions and outcomes.

Algorithms: Set of instructions that a computer or AI system follows to solve a problem or make a decision.

Adaptive leadership: Is a practical leadership and management framework to address, adapt, and thrive in today is VUCA (Volatility, Uncertainty, Complexity, and Ambiguity) business environment. Adaptive leadership is based on flexibility and risk management.

Annex SL: High level structure, provides identical core text, and provides key definitions that will be found in future and revised management system standards.

Annual Risk Report: Document compiled by the Risk Board with consolidated ISO 31000 ERM analysis, risk reports, and treatment plans.

Application controls: Application controls refer to transaction processing controls, sometimes called 'input – processing - output' controls in an IT environment.

Glossary

Assessment risk: Risk the organization did not identify, monitor, or root cause eliminate.

Audit report: Independent review of operations and/or finance for adherence to standards.

Auditability: Ease, consistency, and accuracy of auditing to ISO 9001:2015 or other management system requirements.

Baldrige: Stands for Malcolm Baldrige National Quality Award.

Bias: Intentional or unintentional prejudice from AI systems.

Black Swan: Event that is high consequence and low likelihood.

Board of Directors Audit Committee: Board responsible for reviewing Internal Control over Financial Reporting and operational risks.

Bowtie Method: Risk assessment visual method for looking at potential causes of failure or risk and developing plausible scenario. The reason it is called a bowtie is because the diagrams look like a bow where the causes are on the left side and the consequences are on right side.

Brainstorming: Risk assessment and group problem solving technique to increase the quantity, quality, and diversity of creative ideas

Business assurance: New offering by Certification Bodies to provide higher levels of assurance beyond ISO conformance.

Business case: Rationale for a new project or process.

Business impact analysis: Systematic approach to look at the potential consequences of an interruption in a critical process, business project, disaster, or accident.

Business management system: Set of interrelated generic processes within an organization that focuses on meeting business objectives.

Cause and Consequence Analysis: Risk assessment technique for assessing a chain of consequences. The purpose of the risk assessment is to recognize a series of consequences that originate from a failure, hazard, risk or unexpected events.

Cause and Effect Analysis: Risk assessment technique to analyze the causes of an activity or failure. Cause and effect is also called an Ishikawa or fish bone diagram.

CRO: Acronym for Chief Risk Officer.

Certification Body: Independent company that audits and assures an ISO Quality Management System adheres to requirements of the ISO 9001:2015 standard.

CERM: Acronym for Certified Enterprise Risk Manager® certificate.

Glossary

Checklist: Consists of a series of critical questions (often yes/no) to consider in a risk assessment, process, or activity. Checklist ensures that most critical issues are addressed.

Communication: Process of sharing and getting risk information with stakeholders. Information relates to the existence, extent, management, control, and assurance of risk.

Competence: Ability to apply knowledge, ability, and skills to conduct the work required.

Competency framework: Model for new quality organization. Expression was coined by UK Chartered Quality Institute.

Conformity: Also referred as a conformance. Binary decision (yes/no) to determine adherence to requirements.

Control: Strategy, tactic, or activity that is modifying a risk.

Core functions: Central services and processes of the organization.

Consequence: Outcome of an event that can impact business or other objectives. Event can lead to range of circumstances. Circumstance can be certain or uncertain and can have positive or negative effects.

Consequence rating: Critical rating element and vector of risk along with 'risk consequence'; risk likelihood starts at a 1 rating, which is insignificant to a 5 rating, which is catastrophic.

Context: Environment in which the organization operates and achieves its business objectives.

Continual improvement: Process of surpassing business objectives.

Control: Modifying or changing risk. Controls can be process, policy, device, practice, procedure, or guideline to modify risk. Control may not result in modifying the effect. Often referred to risk - control.

Control environment: Includes: culture, governance, risk management, values, operating style, ethics, and ethos. Sometimes, the control environment is distilled into the expression 'Tone at the Top.'

Control materiality: Reference point to categorize the magnitude of an impact or consequence.

Control owner: Person or function responsible for managing risk and developing the risk policy, procedure, or work instruction.

Controlled risk: Level of risk considering the controls in place.

Corporate knowledge: History of risks provide insight into future threats or opportunities.

COSO: Acronym for Committee of Sponsoring Organizations of the Treadway Commission. COSO is a joint initiative of financial organizations to provide guidance on ERM, GRC, ethics, and financial reporting.

Glossary

COSO ERM: Risk management framework consisting of eight elements: 1. Internal environment; 2. Objective setting; 3. Event identification; 4. Risk assessment; 5. Risk response; 6. Control activities; 7. Information communication; 8. Monitoring.

Corrective action: Process or action to detect and eliminate nonconformance.

CQO: Acronym for Chief Quality Officer.

CSR: Acronym for Corporate Social Responsibility. Usually based on ISO 26000 standard.

Customer: Person (s) who receives a product or service through a value exchange.

Data governance: Managing data used in autonomous decision making systems so it is used responsibly.

Decision Tree Analysis: Risk assessment graphical technique to review decision flow in a tree diagram.

Decision traps: Barriers to RBT, Risk based, problem solving, and Risk based, decision making.

Delphi Method: Structured risk assessment and forecasting technique that relies on a panel of knowledge domain experts to frame and solve a problem.

DIS: Acronym for Draft International Standard.

Disruptive innovation: Innovation that helps create a new market and value that eventually disrupts an existing market and value network.

Documented information: Controlled information to document a QMS or other management system.

Documented policies: Policies outline how the organization meets the requirements of the ISO standard.

Downside risk: Risk usually of negative consequences.

Due diligence: Effort a party makes to avoid harm to another party.

Due care: Care that a reasonable person would exercise under the circumstances; the standard for determining legal duty.

Due professional care: Application of auditing diligence and judgment.

Improved Risk Management: Equivalent Enterprise Risk Management term used in ISO 31000.

Economic consequence: Effect of an occurrence or event on the value of property, process, or facility.

Employee involvement: Employees are aware of the management system and it is applied to their work.

Glossary

Enterprise Risk Management (COSO): Integrated COSO framework published in 2004 defines ERM as a 'process, affected by an entity's Board of Directors, management and other personnel, applied in a strategy setting and across the enterprise, designed to identify potential events that may affect the entity, and manage risk to be within its risk appetite, to provide reasonable assurance regarding the achievement of entity objectives." Comprehensive risk program designed to continuously identify and manage real and potential threats, hazards, and opportunities in the organization. Comprehensive and entity level approach to risk management that engages systems, processes, and activities to improve the quality of Risk based, problem solving, and Risk based, decision making to foster the organization's ability to reach its strategic objectives.

Environmental Risk Assessment: Proactive and systematic process for anticipating and protecting risks to human health, welfare, safety, and environment.

Effectiveness: Ability to meet a desired result or business result.

Efficiency: Being able to meet a desired result using optimized resources.

Emerging risk: Evolving or new risk that is difficult to control or manage since its likelihood, consequence, timing, dependencies, interdependencies, and timing are highly uncertain.

ERM control cycle: Continuing and systematic cycle by which risk is identified; evaluated; risk appetite is determined; risk limits are set; risk is treated; risk is monitored; and risk is assured. Control cycle is often based on a risk management framework cycle.

ERM heavy: Refers to COSO risk management framework.

ERM light: Refers to ISO 31000 risk management framework.

ERM risk map: Graphical representation or roll up of risks for the organization considering current level of controls.

Establishing the context: Internal and external boundary conditions and scope related to managing risk, setting criteria, and defining risk management policy. Definition of the external and internal parameters and conditions to be considered when managing risks. Setting up the risk criteria for the risk management policy.

Event: Occurrence or change in a set of circumstances. Event can have one or more causes. Event is also called an incident or accident. Event with no consequences is called an incident or near miss. Incident and situation that occurs at a specific place during a specific time.

Event tree: Graphical risk assessment used to illustrate the range of probabilities of possible outcomes that can arise from an initiating event.

Event Tree Analysis: Forward looking, bottom-up risk assessment technique that evaluates possible risks. Event tree analyzes the effects of an operating system given that an event has occurred.

Evidence based approach: Rational method for reaching reliable and reproducible audit conclusions and findings.

Glossary

Executive management: Senior organizational management that establishes and review organizational strategic direction; develops strategic risk plan; develops Key Performance Indicators (KPI's) and Key Risk Indicators (KRI's); establishes and embeds risk management culture; oversees risk management; and reviews and approves risk - controls and treatment.

Exposure: Extent to which an organization is exposed to an event.

External context: External environment in which the enterprise operates and establishes its business objectives. External context can include culture, social, political, legal, regulatory, financial, technological, economic, natural, and competitive criteria.

Failure Modes and Effects Analysis: Systematic approach for identifying possible failures modes, which are the ways a product or process may fail. Failure may mean nonconforming products, errors, software defects, or processes not meeting specifications.

Fault tree: Graphic risk assessment tool used to illustrate the range, probability, and interaction of causal occurrences or events that can lead to an erroneous outcome.

Fault Tree Analysis: Used in safety and reliability problem solving. Fault tree analysis includes deductive problem solving to understand the consequences of an event. Fault tree analysis model works backward to understand and deduce what causes and event.

FMEA: Acronym for Failure Mode Effects Analysis.

218 Trust Me: ISO 42001 AI Management System

GAGAS: Acronym for Generally Accepted Government Auditing Standards; professional standards presented in the 2011 revision of Government Auditing Standards provide a framework for performing high quality audits.

General controls: IT controls often described in two categories: IT general controls (ITGC) and IT application controls; used in IT environment; controls, other than application controls, which relate to the environment within which computer based application systems are developed, maintained and operated, and which are therefore applicable to all applications.

Governance: Process by which the Board of Directors reviews the decisions and actions of executive management.

Governance control: Usually a Board level control, which includes oversight, monitoring, and determination of risk appetite.

Governance framework: Roles, responsibilities, and authorities for managing AI responsibly within an organization.

Governance Risk Compliance (GRC): Governance, risk management, and compliance or GRC is the umbrella term covering an organization's approach across areas; while interpreted differently in various organizations. GRC typically encompasses activities such as corporate governance, ERM, and corporate compliance with applicable laws and regulations.

Guide: Facilitator chosen by the auditee to support the audit team.

Glossary

Guide 73: Common risk management definitions that will be incorporated into each management system using Risk Based Thinking (RBT).

Hazard: Man-made source or cause of difficulty or harm, which may be intentional or unintentional.

Hazard Analysis and Critical Control Points (HACCP): Risk assessment approach to assess hazards in manufacturing, food safety, and other production processes. Purpose is to identify hazards occurring in the process and to evaluate control effectiveness at critical control points.

Hazard and Operability Analysis (HAZOP): HAZOP analysis evaluates safety, operating, maintenance, and design risks.

Human Reliability Analysis: Risk assessment technique and study of human factors and human performance used in military, medicine, and manufacturing.

IIA: Acronym for Institute for Internal Auditing.

Impact: Same as consequence. Estimated result including financial and operational that would be realized if a risk event would occur.

Independence: Independence from parties whose risks might be harmed by the results of an audit; specific internal management issues are inadequate risk management, inadequate internal controls, and poor governance.

Information security: Domain that secures the confidential, integrity, and availability of information.

Inherent risk: Risk that the account, disclosure, or financial statement being attested to by an independent firm is materially misstated without considering internal controls, error, or fraud.

Integrated management system: Single management system that integrates elements of multiple management systems.

Integrity: Basis of professionalism. Includes elements of honesty, diligence, responsibility, and honesty.

Intentional hazard: Source of harm or difficulty created by planned course of action.

Interested parties: Person or organization that can impact or be impacted by a decision. Common interested parties may include stakeholders, customers, owners, employees, suppliers, NGOs, regulators, etc.

Internal context: Internal environment in which the enterprise operates and establishes its business objectives. Internal context can include governance, organizational structure, roles, accountabilities, policies, procedures, strategies, plans, tactics, capabilities, resources, perceptions, values, stakeholders, IT, relationships, standards, specifications, contracts, and culture.

Internal control: Integral component of an organization is management that provides reasonable assurance that the following objectives are being achieved: effectiveness and efficiency of

Glossary 221

operations, reliability of financial reporting, and compliance with applicable laws and regulations.

Internal review: Review undertaken to assess the suitability, adequacy, effectiveness, and efficiency of operational systems and to determine opportunities for improvement.

Intervention risk: Risk the organization did not treat and correct the problem at the symptom and root cause levels.

Issues log: Record of issues faced and actions taken to remediate them. Issues identified as risks are assessed and treated.

ISO: Acronym for International Organization of Standardization.

ISO 9001:2015: ISO Quality Management System standard, which was finalized in Q3 of 2015.

ISO 19011: Comprehensive standard for managing, planning, conducting, and reporting management system certification audits.

ISO 31000: Risk management framework or guideline used as reference for many ISO families of standards.

ISO 42001: Artificial intelligence management systems standard.

Key Performance Indicator (KPI): Measure (s) on the achievement toward a control objective.

KRI: Acronym for Key Risk Indicator. Measure (s) that can provide an early warning that a risk has occurred or is recurring.

Layer of Protection Analysis: Risk assessment technique is a process risk assessment and hazard tool, which looks at potential hazardous events, their consequences, initiating causes, and likelihoods of occurring.

Level of risk: Magnitude of risk. Expressed in terms of likelihood and consequence.

Likelihood: Possibility or chance of something occurring or happening. Likelihood is also called probability or frequency.

Management commitment: Senior management is committed to the planning, implementation and maintenance of the management system.

Management control: Controlling is one of the managerial functions like planning, organizing, staffing and directing; important function because it helps to check the errors and to take the corrective action so that deviation from standards are minimized and stated goals are achieved. How well the organization is identifying, controlling, and mitigating risk.

Management system: Interrelated processes within an organization whose aim is to achieve business objectives.

Markov Analysis: Risk assessment method to evaluate the relative reliability of system components. Can be used to determine dependencies between components, personnel, and technologies.

Glossary

Materiality: Concept in accounting and auditing relating to the importance or significance of an amount, transaction, finding, or discrepancy.

Materiality levels: Thresholds the organization uses to determine risks at the enterprise and entity level.

Measurement: Process to determine a value or number.

Mitigation plan: Strategy for risk mitigation. If an identified risk is not within the risk appetite, risk tolerance, or risk retention, then further mitigation is planned.

Monitoring: Continual checking, observing, and supervising the status of risk to determine changes that may affect controls or residual risk.

Monte Carlo Method: Risk assessment method for analyzing engineering and physical phenomena. Monte Carlo methods are used for optimization. Monte Carlo method can be used with a probabilistic distribution so information can be inferred from a large population of data.

Natural hazard: Source of harm created by environmental phenomena.

Nonconformity: Also referred as a nonconformance. Binary decision (yes/no) to determine adherence to requirements. In this case, an inability to conform to requirements.

Objective: Result to be gained or achieved.

Observer: Person who accompanies the audit team and ensures that procedures are followed.

Operational risk: Potential of loss attributable to process variation or disruption in operations caused by internal or external factors.

Opportunity: Chance for an advancement, improvement, or progress.

Organization: Legal entity that has processes and functions that achieve business objectives.

Outsource: External organization providing products or services.

PDCA: Acronym for Plan – Do – Check – Act.

PESTLE: Acronym for Political, Economic, Sociological, Technical, Legal, and Environmental. Commonly used as a planning tool to identify and prioritize threats in the external environment.

People risk: People are the backbone and personality of a business; people are also a key source of risk, because risk management is the fundamental driver to sustainable success, understanding the various risks from employees that must be a top priority for business leaders and policymakers.

Performance: Actionable or measurable output or outcome.

Personally identifiable information: Information that can be used to identify a person directly or indirectly.

Glossary

Preliminary Hazard Analysis: Preliminary hazard analysis is the review of potential threats, events, or risks. Hazard is a potential condition that may exist or not occur. Hazard may not be anticipated or planned. Hazard may be unknown or even unknowable due to a potential 'black swan', cascading risk, or interactive risk factors.

Preventive action: Change implemented to address a weakness in a management system that is not yet responsible for causing nonconforming product or service.

Priority risk: Risks that are high after risk - controls and risk mitigation/treatment.

Privacy: Protecting personal individual information in AI systems, respecting privacy rights, and complying with regulations.

Probability: Measure of the change of occurrence, usually a number between 0 and 1 where 1 is absolute certainty.

Process: Interrelated and/or interacting activities that add value to inputs to create an output.

Process analysis: Approach to evaluate operational and service performance and identify opportunities for improvement.

Process risk: Probability of loss inherent in a business process; may include lack of process capability, lack of process stability, and/or lack of improvement.

Professional judgment: Standard of care that requires auditors to exercise reasonable care and diligence and to observe the principles of serving the public interest and maintaining the highest degree of integrity, objectivity, and independence in applying professional judgment to all aspects of their work.

Project risk: Project risk involves not being able to meet project objectives or deliverables based on project scope, quality, schedule, or cost.

Quality assurance: Engineering activities implemented in a quality system so product or service requirements can be fulfilled.

Quality management: Management of quality related activities, including assurance and control.

Quality Management Thinking (QMT): Quality equivalent to Risk Based Thinking (RBT).

Quality governance: Process by which the Board of Directors reviews RBT, quality decisions and actions of executive management.

QMS: Acronym for Quality Management System.

RBT: Acronym for ISO Risk Based Thinking. First stage in a Risk Capability Maturity Model (RCMM) journey from RBT to risk assessment, risk management, to ISO 31000 ERM.

RCMM: Acronym for Risk Capability Maturity Model, which consists of five levels from ad hoc to optimized.

Reasonable assurance: Most cost-effective measures are taken in the design and implementation stages to reduce risk and restrict expected deviations to a tolerable level.

Red Book: Institute of Internal Auditors (IIA) guidelines for conducting an internal audit.

Reliability Centered Analysis: Risk assessment analysis focuses on long term quality and lifecycle management of a product. Can be used in reliability centered maintenance, product failure, or operational safety analysis.

Reputational risk: Decrease in brand equity or credibility in the organization.

Requirement: Explicit or implicit needs or expectations. Customers, regulators, or interested parties can develop a requirement.

Residual risk. Risk remaining after risk treatment. Residual risk can contain unidentified risk. Residual risk is also called retained risk. Exposure to loss remaining after other known risks have been countered, factored in, or eliminated.

Review: Activity used to determine the suitability, adequacy, and effectiveness of risk - controls against established objectives.

Risk: Uncertainty on achieving a business objective. Risk is also a deviation from an objective, which can be either positive or negative. Objectives can be from the financial, quality, project, process, program, transactional, or supply chain. Qualitative risk is defined as likelihood and consequence. Potential that a chosen action or activity (including the choice of inaction) will lead to a loss (an undesirable outcome). Qualitative risk is defined as the consequence and likelihood of an event. Potential event with a negative or undesirable outcome, which may include the potential failure to capitalize on an opportunity.

Risk analysis: Process to understand the nature of risk and to determine the level of risk. Risk analysis is the basis for risk evaluation and decisions about risk treatment and risk management.

Risk acceptance: Explicit or implicit risk based decision to take no action that would impact or affect all or part of a specific risk.

Risk aggregation: Collection of risk (categories and impact) to develop an understanding of the risk to the enterprise.

Risk analysis: Systematic examination of risk components and characteristics.

Risk appetite: Level of risk that an organization is prepared to accept before action is deemed necessary to reduce it; also sometimes called 'risk tolerance.' Amount and type of strategic risk the organization is willing to pursue and manage.

Risk assessment: Process of identifying, analyzing, and evaluating risk. Determination of quantitative or qualitative value of risk related to a concrete situation and a recognized threat (also called hazard).

Glossary

Risk assurance: Ability to provide requisite level of risk - control effectiveness and confidence of effectiveness controls.

Risk attitude: Organization's approach to assess and mitigate risk. Risk management is an element of risk attitude.

Risk aversion: Attitude and policy to move away and not pursue opportunities and actions.

Risk avoidance: Risk based, decision making to be not involved in an activity or to withdraw from an activity in order not to be exposed to a specific risk.

Risk Based Auditing: Also known as Value Added Auditing. Red Book and Yellow Book are examples of Risk Based Auditing.

Risk Based Certification: Det Norske Veritas (Certification Body(term for high level, risk certification and assurance.

Risk based, decision making. Critical element of ISO Risk Based Thinking (RBT).

Risk based, problem solving. Critical element of ISO Risk Based Thinking (RBT).

Risk Based Thinking (RBT): International Organization for Standardization (ISO) tagline for ISO 9001:2015 and possibly families of standards as it incorporates risk. ISO says that RBT has been part of ISO standards. RBT is defined in this book as: 1. Risk based, problem solving and 2. Risk based, decision making.

Risk Based Thinking journey: Steps in risk journey from RBT to risk assessment, risk management, to ISO 31000 ERM.

Risk Capability Maturity Model (RCMM): Methodology used to define and identify an organization's risk - control processes, procedures, and protocols. Model is often based in terms on a five-level evolutionary scale from ad hoc to world class, risk management.

Risk category: Distinct classes of risk, where similar opportunities and risks can be analyzed and compared.

Risk - control: Method by which firms evaluate potential losses and take action to reduce or eliminate such threats. Risk - control is a technique that utilizes findings from risk assessments (identifying potential risk factors in a firm is operations).

Risk criteria: Terms against which risk is evaluated. Risk criteria are based on business objectives, external/internal context, and other criteria. Risk criteria can be based on standards, laws, policies, and other requirements.

Risk description: Structured statement of risk containing five elements: source of risk, events, causes, consequences, and likelihood.

Risk escalation: Communication of risks to the appropriate level of management requiring additional resources for treatment.

Risk evaluation: Process of comparing the results of risk analysis against risk criteria to determine whether it is acceptable or tolerable to the enterprise. Risk evaluation is used in the decision of risk treatment. Assessing probability and impact of personal risks considering any interdependencies or other factors outside the immediate scope under investigation.

Risk event: Occurrence or change in a particular set of circumstances that usually have negative consequences.

Risk governance: Processes to ensure authorities and accountabilities for managing enterprise risk, deploying risk management framework, implementing risk management process, and proving risk assurance is appropriate to the organization.

Risk identification: Process of finding, recognizing, and describing risk. Risk identification involves identifying risk sources, events, likelihood, and possible consequences. Risk identification involves historical data, theoretical analysis, expert opinions, and stakeholder needs.

Risk inventory: List of prioritized organizational risks.

Risk level: Nature and threshold for risk based on likelihood and consequence assessment.

Risk likelihood rating: Critical rating element and vector of risk along with 'risk consequence'; risk likelihood starts at a 1 rating, which is a rare event to a 5 rating, which is almost certain.

Risk limit: Threshold used to monitor actual risk exposure of a unit or units of an organization to ensure the level of aggregate risk remains within the risk appetite or risk tolerance.

Risk management: Identification, assessment, and prioritization of risks (effect of uncertainty on objectives, whether positive or negative) followed by effective and economic application of resources to minimize, monitor, control, and assure the probability and/or consequence of negative events or to maximize opportunities.

Risk management framework: Process cycle for managing risk. ISO 31000 and COSO are two common risk management frameworks.

Risk management plan: Steps, procedures, approach, resources, methodology, and components applied to the management of upside and downside risk.

Risk management: Also, called risk mitigation or risk - control. Risk management is usually defined as the control of risk.

Risk management framework: Structure upon which to build strategy or set of controls organized in categories to be able to reach objectives, monitor, and assure performance.

Risk management policy: Policy is the highest-level documentation and organizational direction relating to risk management.

Risk management process: Systematic application of organizational policies, procedures, work instructions, processes, protocols, practices, and guidelines for establishing the context, analyzing, assessing, treating, monitoring and communicating risks.

Glossary

Risk map: Visual method of laying out the risk of an event or variation; visual representation of statistics; usually consisting of red, yellow, and green elements.

Risk Management System: Similar to ISO management systems, such as Quality Management System (QMS) and Environmental Management System (EMS). ISO 31000 can form the basis for a risk management system.

Risk matrix: Risk assessment tool for ranking and illustrating components of risk in an array.

Risk metric: Measure of risk. Can include Key Risk Indicators, value at risk, or tail exposures, Cpk etc.

Risk mitigation: System, process, or investment to control the likelihood or consequence of a risk.

Risk monitoring: Last major element of ISO 31000 risk management process, used to determine if risk management plan is being followed and if internal risk - controls are working effectively.

Risk owner: Enterprise owner with the accountability, authority, and responsibility to manage risk. Additional responsibilities include identify risks in span of control; ensure risks with control area are managed appropriately; develop treatment plans; monitor risk - control and treatment; ensure treatment owners are assigned; place risks on enterprise register; and escalate risks if necessary.

Risk profile: Description of the set of risks that can relate to the enterprise. Comprehensive view of risk the organization faces.

Risk reduction: Also called risk - control.

Risk register: Record of risk information by level and type of risk.

Risk reporting: Form of risk communication intended to inform stakeholders, customers, or interested parties' current state of risk and its treatment.

Risk response: Used similarly as risk treatment or risk mitigation; appropriate steps taken or procedures implemented upon discovery of an unacceptably high degree of exposure to one or more risks.

Risk retention: If a risk is within the organizational risk appetite, then the risk is accepted; no additional controls are required but are continuously monitored for suitability.

Risk sharing: Form of Risk based, decision making involving the agreed upon distribution of risk among parties.

Risk source: Element (s) that have the potential that can cause risk.

Risk syntax: Use and application of risk concepts. See 'risk taxonomy.'

Risk taxonomy: Practice and science of risk classification of things or concepts and principles that underlie the classification.

Glossary

Risk tolerance: Acceptable level of variation a company or an individual is willing to accept in the pursuit of the specific objective. Aggregate risk-taking capacity of the enterprise.

Risk transfer: Form of Risk based, decision making to manage risk so that it shifts some or all the risk to another party, system, process, geography, supplier, or network.

Risk treatment: Process of managing risk, including searching for opportunities; avoiding risk; increasing risk; removing the risk source; changing likelihood; changing consequences; sharing the risk; or retaining the risk. Risk treatment is called risk elimination, risk prevention, risk reduction, risk response, or risk management.

Root cause analysis: Risk based, problem solving method to identify the primary cause of a recurring, chronic, or systemic problem.

Security: Ensuring AI systems are not penetrable and resilient to attacks.

Scenario analysis: Risk assessment process to identify, assess, and evaluate possible outcomes based upon specific assumptions. Outcomes are potential and realizable projections or alternatives of the future.

Scenario test: Process for assessing the adverse consequences of one or more possible events occurring simultaneously or serially.

SMART objectives: Acronym representing: Specific, Measurable, Assignable, Realistic, and Timely

Sneak circuit analysis: Risk assessment method to determine safety of mission critical electronic and mechanical components.

Self-certification: Statement by an organization that it meets ISO requirements without a third party, certification audit.

Speed of onset: Time it takes for a risk event to occur or manifest.

Stakeholder: Enterprise or person that can impact or be impacted by risk.

Stakeholder analysis: Process identifying individuals who have a vested interest in achieving business objectives and uncertainties.

Stakeholder engagement: Users, interested parties, and customers who are involved in the architecting, designing, and deploying AI systems.

Stress test: Process for measuring the adverse consequence on one or more quality, supplier, ISO, design, cyber security, information technology, people, or other operational factors that can impact the organization's financial profile.

Structured interview: Follows a standardized checklist or procedure to conduct a series of interviews. Structured interview is used as part of other types of quantitative or qualitative risk assessments or survey research to scope or frame a problem.

Structured 'What IF' Technique: Also called SWIFT technique. Swift is a high to low risk assessment technique that is often used

Glossary

with other risk assessment tools such as Failure Mode and Effects Analysis and brainstorming.

Supplier: Organization or person providing a product or service as part of a value exchange.

Survey: Gather data on risk, threats, and hazards.

SWOT: Acronym for Strengths – Weaknesses – Opportunities – Threats. Commonly used as a planning tool for assessing risk and evaluating a business.

System: Set of interrelated and interacting activities.

Technical Committee (TC) 176: ISO committee responsible for writing QMS standard, ISO 9001:2015.

Technical controls: Example of control usually with IT and ICS systems; control consists of identification and authentication.

Technical expert: Person with specific knowledge who is part of the audit team.

Technology risk: Risk that key technology processes a company uses to develop, deliver, and manage its products, services, and support operations do not meet requirements.

Threat: Natural or man-made activity with the potential to cause damage, injury, or loss.

TQM: Acronym for Total Quality Management. Implies a high level of quality maturity and capability.

Treatment owner: Person responsible for treating the risk.

Upside risk: Opportunity or positive risk.

VUCA: Acronym for Volatility, Uncertainty, Complexity, and Ambiguity. Description of current business environment that requires different strategic and tactical planning models.

Vulnerability: Susceptibility of the enterprise to a risk event related to the entity's preparedness, agility, and adaptability.

White space risk: Risks in the white spaces between silos, functions, and work between processes.

Yellow Book: Also called GAGAS or Generally Accepted Government Auditing Standards. Audit standards for public auditing in U.S. and Canada.

Index

A

AI, 1, 3, 5, 7, 8, 9, 10, 11, 12, 13, 15, 16, 17, 18, 19, 20, 21, 22, 23, 24, 25, 26, 27, 28, 29, 31, 32, 33, 34, 35, 36, 37, 38, 39, 40, 41, 42, 44, 45, 47, 48, 49, 50, 51, 52, 53, 54, 55, 56, 57, 58, 59, 60, 61, 63, 65, 66, 67, 68, 70, 71, 72, 73, 74, 75, 76, 77, 78, 79, 80, 81, 83, 84, 85, 86, 87, 88, 89, 90, 91, 92, 93, 94, 95, 96, 97, 98, 99, 100, 101, 102, 103, 104, 105, 106, 107, 109, 110, 111, 112, 113, 114, 115, 116, 117, 118, 119, 121, 122, 123, 124, 125, 126, 127, 129, 130, 131, 132, 133, 134, 135,136, 137, 138, 139, 140, 141, 142, 143, 144, 147, 148, 149, 155, 156, 157, 158, 159, 160, 161, 162, 163, 164, 165, 166, 167, 168, 169, 170, 171, 172, 173, 174, 175, 176, 177, 178, 179, 180, 181, 182, 183, 184, 185, 186, 187, 188, 189, 190, 191, 192, 193, 194, 195, 196, 197, 198, 199, 200, 201, 202, 203, 204, 205, 207, 208, 209, 218, 225, 235, 236, 239
AI assurance, 35, 98, 141, 143
AI audit, 71, 72, 73, 134, 135
AI control objective, 204
AI effectiveness, 93, 140, 141
AI governing body, 91, 92, 94, 113
AI guidelines, 45, 81, 95, 113, 158, 174, 198
AI information security, 38, 65, 73, 76, 103, 115, 122, 133, 134
AI management system, 37, 64, 66, 67, 68, 83, 88, 89, 91, 92, 93, 95, 97, 105, 106, 107, 126, 129, 134, 135, 136, 137, 142, 207
AI measurement, 130, 131, 137
AI monitoring, 67, 94, 95, 126, 129, 131, 132, 134, 135, 137, 141, 182, 183, 198
AI nondormancies, 141
AI objectives, 17, 21, 22, 63, 64, 65, 89, 92, 93, 95, 100, 103, 104, 105, 107, 111, 115, 129, 132, 136, 140, 141, 147, 157, 174, 179
AI policies, 49, 63, 84, 93, 94, 100, 110, 113, 114,

115, 116, 155, 157, 158, 160, 170, 196
AI process, 43, 44, 71, 73, 100, 105, 107, 116, 123, 129, 130, 131, 134, 135, 138, 139, 157, 158, 164, 169, 188, 189
AI process maps, 116
AI regulations, 47, 92
AI requirements, 37, 38, 64, 76, 88, 89, 93, 111, 116, 118, 121, 122, 141, 160, 175, 176, 178, 179, 180, 187, 189, 203
AI risk appetite, 124
AI risk management, 1
AI risk treatment, 125
AI risks, 21, 22, 27, 100, 101, 102, 124, 125
AI system, 11, 19, 23, 26, 28, 34, 37, 39, 42, 48, 49, 50, 54, 55, 56, 57, 58, 59, 60, 67, 68, 74, 77, 78, 87, 88, 89, 93, 98, 100, 101, 102, 103, 113, 118, 119, 122, 125, 126, 131, 132, 138, 141, 142, 144, 148, 157, 161, 162, 163, 164, 165, 166, 167, 168, 169, 170, 174, 176, 178, 179, 180, 181, 182, 183, 185, 186, 187, 188, 195, 198, 203, 205, 207, 208
AI system life cycle, 173, 175
AI technology, 122
AI trust, 50

AIMS, 41, 43, 44, 63, 66, 67, 68, 70, 71, 72, 74, 75, 76, 77, 81, 83, 84, 86, 87, 89, 90, 91, 92, 93, 94, 95, 96, 97, 98, 103, 104, 106, 107, 109, 110, 111, 112, 113, 114, 117, 118, 119, 121, 122, 126, 130, 131, 136, 137, 138, 142, 143, 144, 148, 149, 150, 155, 157, 158, 161, 163, 164, 166, 167, 170, 174, 175, 181, 185, 192, 204, 205, 207
Annex A, 70, 101, 148, 149
Annex B, 70, 149
Annex C, 70, 149
Annex D, 70, 149
Annexes, 8, 70, 72, 74, 101, 147, 150
assessing impacts of AI systems, 151
assurance, 71
audit, 36, 39, 40, 41, 42, 43, 71, 118, 129, 134, 135, 171, 199, 205, 216, 218, 219, 224, 227, 236, 237
auditing, 9, 39, 40, 72, 218, 219, 229, 238

B

benchmarks, 99
biases, 113, 130, 131, 143
black box texting, 132
business and work automation, 19

Index

business impact assessment, 126

C

certification, 39, 40, 67, 71, 74, 75, 78, 221, 229, 236
Certification Body, 39, 118, 210
compliance, 35, 40, 41, 42, 47, 48, 50, 52, 58, 59, 71, 73, 74, 75, 77, 79, 115, 116, 117, 133, 141, 144, 148, 171, 188, 193, 197, 198, 199, 201, 218, 221
conformity assessment, 36, 37, 38, 40, 48, 76
context, 58, 70, 83, 84, 85, 89, 211
continual AI improvement, 139, 140
continuous improvement, 78, 109
control objectives, 155, 156, 157, 158, 159, 160, 161, 162, 163, 164, 165, 166, 167, 168, 169, 170, 171, 172, 174, 175, 177, 179, 180, 181, 182, 183, 184, 185, 186, 187, 188, 189, 190, 191, 192, 193, 194, 195, 196, 197, 198, 199, 200, 201, 202, 203
controls, 8, 44, 59, 70, 74, 75, 77, 101, 102, 125, 130, 143, 148, 149, 156, 157, 158, 160, 161, 163, 164, 165, 166, 167, 169, 170, 171, 172, 174, 175, 176, 178, 179, 181, 182, 183, 185, 186, 187, 188, 189, 191, 192, 193, 194, 195, 197, 198, 199, 201, 202, 203, 205, 207, 208, 212, 216, 218, 219, 220, 223, 229, 232, 234, 237, 238
corrective action, 64, 65, 73, 132, 142, 222
criminal justice, 19
CSF, 33

D

data for AI system, 152
data testing, 131
decision making risks, 23
decision making, 3
deployment, 77, 156, 157, 158, 160, 161, 163, 164, 165, 166, 167, 169, 170, 171, 172, 174, 175, 176, 177, 178, 179, 180, 181, 182, 183, 185, 186, 187, 188, 189, 191, 192, 193, 194, 195, 197, 198, 199, 201, 202, 203
documentation, 51, 67, 106, 113, 115, 116, 186, 191, 198, 232

E

employment screening, 19
EU AI Act, 47, 50, 54, 92

executive management, 74, 75, 85, 91, 92, 93, 94, 95, 218, 226, 238
existential risk, 20
explainability, 50, 68, 136, 171, 185, 207
external context, 83, 217

F

facial recognition, 19
financial institutions, 53
financial services legislation, 53
fit for purpose, 38, 42, 77, 164, 187, 198, 199
framework, 27, 33, 52, 66, 68, 79, 80, 93, 208, 211, 212, 215, 216, 218, 221, 231, 232

G

governance, 84, 117, 144, 218
governing body, 91, 92
harmonised standards, 49, 53

H

healthcare access, 19
high risk, 35, 37, 48, 52, 59, 205
high AI products, 95, 112, 115
homelessness, 20

human, 2, 7, 10, 12, 15, 16, 18, 19, 24, 25, 26, 28, 29, 30, 31, 55, 66, 68, 72, 79, 103, 109, 111, 127, 156, 167, 171, 176, 178, 215, 219
human competence, 111
human in the loop, 25
human safety, 34

I

information assets, 76, 122, 133, 134
information for interested parties, 153
inspection, 40
intended purpose, 55, 58, 59, 60, 180, 182, 187
interested party, 64, 71, 73, 76, 83, 87, 88, 95, 104, 112, 115, 118, 122, 123, 137, 140, 203
internal audit, 134
Internal context, 83, 84, 220
internal organization, 150, 151
International Organization for Standardization, 78, 229
ISO 27001, 38, 65, 72, 99, 204
ISO 31000, 21, 22, 54, 208, 214, 216, 221, 226, 230, 232, 233
ISO 42001, 1, 3, 5, 7, 8, 9, 10, 12, 13, 21, 26, 28, 32, 35, 38, 39, 41, 42, 43, 44,

Index

54, 61, 63, 65, 66, 67, 68, 69, 70, 71, 72, 73, 74, 75, 76, 77, 78, 79, 80, 81, 83, 85, 86, 89, 90, 91, 93, 96, 97, 98, 99, 100, 101, 102, 103, 104, 105, 107, 109, 111, 112, 114, 115, 116, 117, 118, 119, 121, 122, 123, 129, 130, 131, 133, 138, 139, 143, 144, 147, 148, 149, 150, 160, 198, 204, 221
ISO 9001, 37, 47, 48, 53, 64, 209, 210, 221, 229, 237
ISO management systems, 27, 63, 139, 233
ISO standards, 27, 38, 41, 69, 73, 78, 92, 97, 111, 116, 119, 122, 123, 130, 132, 133, 136, 140, 229

L

lifecycle approach, 74, 76, 77, 204
limited risk, 36

M

machine learning, 189
management assessment, 137
management system, 8, 9, 26, 38, 41, 42, 44, 47, 48, 53, 54, 63, 64, 65, 66, 67, 70, 71, 72, 73, 78, 80, 83, 91, 111, 115, 129, 143, 175, 208, 209, 210, 214, 219, 220, 221, 222, 225
management systems standards, 63
managing documentation, 106
model development, 163, 165, 193
model testing, 132

N

NIST, 33, 45, 72, 75, 239
nonconformances, 123
nonconformities, 141

O

objectives, 17, 23, 39, 41, 44, 69, 70, 72, 74, 77, 79, 90, 95, 98, 99, 100, 103, 104, 105, 109, 113, 121, 122, 123, 124, 130, 140, 147, 149, 150, 155, 158, 160, 174, 175, 177, 193, 198, 199, 201, 204, 205, 210, 211, 212, 215, 217, 220, 222, 224, 226, 227, 230, 232, 235, 236
operational planning, 121, 124
opportunities, 27, 68, 71, 80, 92, 97, 98, 101, 107, 111, 126, 137, 139, 212, 215, 221, 225, 229, 230, 232

P

PaperClip Analogy, 20
performance, 21, 38, 51, 59, 64, 65, 73, 76, 89, 92, 94, 101, 103, 110, 115, 122, 123, 129, 131, 132, 134, 136, 137, 141, 179, 181, 187, 191, 207, 219, 225, 232
performance testing, 132
personally identifiable information, 133, 224
PII, 133
policies related to AI, 150
post-market monitoring system, 51, 55
prevent problems, 132
preventive Actions, 142
probabilistic thresholds, 59
process control, 122
process management, 123
product mark, 40
provenance, 186

Q

quality assurance, 49
quality control, 49
quality management, 49, 226
Quality Management System, 47, 48, 221, 226, 233

R

regulatory compliance, 48
release criteria, 179
reliability, 94, 131, 182, 217, 221, 222, 227
reputation management, 80
residual risk, 57, 223
resource allocation, 141
resources for AI system, 151
responsible execution, 159
risk, 3
risk assessment, 66, 74, 75, 100, 101, 102, 124, 126, 158, 210, 213, 216, 217, 222, 226, 230, 236
risk avoidance, 125, 229
risk based, decision making, 15
risk based, problem solving, 27, 28, 77, 86, 118, 205
risk control, 148, 150, 156, 157, 158, 160, 161, 163, 164, 165, 166, 167, 169, 170, 171, 172, 173, 174, 175, 176, 177, 178, 179, 180, 181, 182, 183, 185, 186, 187, 188, 189, 191, 192, 193, 194, 195, 197, 198, 199, 201, 202, 203
risk control architecture, 150
risk criteria, 51, 98, 99, 121, 216, 230
risk decisions, 27, 87, 92, 113, 126, 132, 136, 137, 169
risk evaluation, 48, 228
risk management, 21, 22, 33, 50, 54, 55, 56, 57, 58, 60, 66, 68, 79, 80, 207,

Index

208, 212, 215, 216, 217, 218, 219, 224, 226, 228, 230, 231, 232, 233, 235
risk management system, 50, 54, 56, 60, 233
risk mitigation, 125, 233
risk transfer, 125, 235
risk treatment, 22, 101, 125, 126, 227, 228, 231, 234
risk based, problem solving, 15
risks, 17, 21, 22, 27, 38, 52, 54, 55, 56, 57, 58, 60, 65, 68, 71, 76, 84, 92, 93, 94, 97, 100, 101, 112, 115, 118, 123, 124, 125, 126, 134, 141, 143, 149, 156, 160, 162, 169, 190, 191, 207, 209, 212, 215, 216, 219, 221, 223, 224, 225, 227, 230, 231, 232, 233, 234
rules of engagement, 22, 23, 24, 26, 28, 29

S

safeguards, 40, 133
scope, 22, 50, 67, 72, 88, 89, 216, 226, 231, 236
self certify, 71, 76
self-declaring, 41
senior leadership, 135
shall, 89
should, 89
SMART, 104, 105, 235
social and technical policies, 158

social criteria, 23
societal impact assessment, 158
stakeholder engagement, 158
Sub Control Objective, 150, 151, 152, 153, 154
suppliers, 202, 220
system owners, 201

T

technical specifications, 49
technical systems, 86, 127
tenant screening, 19
testing, 40, 58, 59, 179
testing procedures, 59
third party inter-relationships, 154
top management, 92, 95, 238
training data, 77, 131, 176, 185
trust, 7, 10, 11, 12, 13, 15, 18, 20, 25, 26, 29, 31, 37, 38, 39, 66, 68, 72, 80, 124, 130, 136, 204, 239
Trust Manifesto, 34
trust me, 1
trustworthiness, 18, 25, 130, 239
trustworthy, 16, 17, 31, 68, 135, 176, 239

U

unacceptable deviations, 130

unacceptable risk, 35
uncertainty, 11, 17, 26, 72, 131, 232
understanability, 192
unintended consequences, 123
use cases, 116

use of AI systems, 153
user feedback, 177

V

VUCAN, 3

Endnotes

[1] MIT Sloan Review, Eric Schmidt, November 2022.
[2] 'Trustworthy AI,' IBM website, https://research.ibm.com/topics/trustworthy-ai, April 2024.
[3] 'AI Trust Gap,' Mitre, https://www.mitre.org/focus-areas/artificial-intelligence/ai-trust-gap, 2024.
[4] Trusted AI 101: A Guide to Building Trustworthy and Ethical AI Systems,' DataRobot, https://www.datarobot.com/trusted-ai-101/April 2024
[5] 'AI and Human Trust in Healthcare: Focus on Clinicians,' JMR, June 2020.
[6] 'Establishing Trust in Generative Ai,' Thomson Reuters, October 9, 2023.
[7] 'NIST Proposes Method for Evaluating User Trust in AI Systems,' NIST website, Brian Stanton, https://www.nist.gov/news-events/news/2021/05/nist-proposes-method-evaluating-user-trust-artificial-intelligence-systems, April, 2024.
[8] 'AI Trust Gap,' Mitre, https://www.mitre.org/focus-areas/artificial-intelligence/ai-trust-gap, 2024.
[9] 'Explainable AI: How Humans Can Trust, AI,' Ericsson, https://www.ericsson.com/en/reports-and-papers/white-papers/explainable-ai--how-humans-can-trust-ai, April, 2024.
[10] How Can We Trust AI If We Don't Know How It Works, Scientific American, Mark Bailey, October 3, 2023,
[11] '90% Staff Replaced With AI Chatbot Lina, Dukaan Founder Suumit Shah Justifies Sacking As 'Tough But Necessary'; Social Media Thinks Otherwise,' July 13, 2023. India Times.
[12] 'Trust in AI, Global Insights 2023', KPMG, February 2023.
[13] 'Trust in AI, Global Insights 2023', KPMG, February 2023.
[14] 'AI and Trust,' Bruce Schneier, Harvard Kennedy School, November, 2023
[15] Contextualizing End User Needs: How To Measure the Trustworthiness of an AI System,' Carrie Gardner, Katherine Marise Robinson, Carol Smith Alexandrea Steiner SEI Blog, https://insights.sei.cmu.edu/blog/contextualizing-end-user-needs-how-to-measure-the-trustworthiness-of-an-ai-system/, July 17, 2023.

[16] How Can We Trust AI If We Don't Know How It Works, Scientific American, Mark Bailey, October 3, 2023,
[17] How Can We Trust AI If We Don't Know How It Works, Scientific American, Mark Bailey, October 3, 2023,
[18] NetApp, https://www.netapp.com/artificial-intelligence/what-is-artificial-intelligence/, 2022.
[19] Quoted in Public Trust in AI I Sinking Across the World, Justin Westcott, Axios, March 5, 2024,
[20] 'How Do You Teach AI the Value of Trust,' Cathey Cobey, Jeanne Boillet, EY, https://www.ey.com/en_gl/insights/digital/how-do-you-teach-ai-the-value-of-trustApril, 2024,
[21] 'Our World: No One Should Trust AI,' November 11, 2018, Joanna Bryson.
[22] 'Trust Manifesto,' ARM AI, 2019.
[23] 'Excellence and Trust In AI,' European Commission, https://commission.europa.eu/strategy-and-policy/priorities-2019-2024/europe-fit-digital-age/excellence-and-trust-artificial-intelligence_en, April 2024.
[24] ISO/IEC 42001 Certification 0 AI Management System, SGS, https://www.sgs.com/en/services/iso-iec-42001-certification-artificial-intelligence-ai-management-system, April , 2024.
[25] 'AI Trust Standard & Label, ' https://oecd.ai/en/catalogue/tools/ai-trust-standard-and-label, April, 2024.
[26] Can We Trust AI?, Mazumdar, Caltech, https://scienceexchange.caltech.edu/topics/artificial-intelligence-research/trustworthy-ai, April, 2024.
[27] Article 17, Quality Management System, paragraph 1, EU AI Act, 2024.
[28] Article 9, Risk Management System, EU AI Act, 2024
[29] 'AI Management Systems: What Business Need To Know, ISO, https://www.iso.org/artificial-intelligence/ai-management-systems, https://www.iso.org/artificial-intelligence/ai-management-systems, April, 2024.
[30] 'Can We Trust AI?, Caltech, https://scienceexchange.caltech.edu/topics/artificial-intelligence-research/trustworthy-ai, April, 2024.
[31] Contextualizing End User Needs: How To Measure the Trustworthiness of an AI System,' Carrie Gardner, Katherine Marise

Robinson, Carol Smith Alexandrea Steiner SEI Blog, https://insights.sei.cmu.edu/blog/contextualizing-end-user-needs-how-to-measure-the-trustworthiness-of-an-ai-system/, July 17, 2023.

[33] 'Our World: No One Should Trust AI,' November 11, 2018, Joanna Bryson.

[34] 'Trustworthy AI,' IBM website, https://research.ibm.com/topics/trustworthy-ai, April 2024.

[35] Contextualizing End User Needs: How To Measure the Trustworthiness of an AI System,' Carrie Gardner, Katherine Marise Robinson, Carol Smith Alexandrea Steiner SEI Blog, https://insights.sei.cmu.edu/blog/contextualizing-end-user-needs-how-to-measure-the-trustworthiness-of-an-ai-system/, July 17, 2023.

[36] 'How Can We Trust AI If We Don't Know How It Works, Scientific American, Mark Bailey, October 3, 2023,

Made in the USA
Columbia, SC
05 June 2024